ESCAPING NAT-URE

How to Survive Global Climate Change

Duke University Press *Durham and London* 2024

ESCAPING NAT-URE

ORRIN H. PILKEY

with Charles O. Pilkey Linda P. Pilkey-Jarvis Norma J. Longo
Keith C. Pilkey Fred B. Dodson Hannah L. Hayes

Chapter-opening photo credits: *Earth*—Charles Pilkey; *Air*—NASA; *Fire*—US Forest Service, USDA / Kari Greer; *Water*—© Gina Longo. Used with permission; *Space*—USDA (digitally remade by Charles Pilkey).

Printed in the United States of America on acid-free paper ∞

Project editor: Lisa Lawley
Designed by Courtney Leigh Richardson
Typeset in Warnock Pro and Work Sans by Copperline Book Services

Library of Congress Cataloging-in-Publication Data
Names: Pilkey, Orrin H., [date] author. | Pilkey, Charles O., author. | Pilkey-Jarvis, Linda, author. | Longo, Norma J., [date] author. | Pilkey, Keith C., [date] author. | Hayes, Hannah L., author. | Dodson, Fred B., author.
Title: Escaping nature : how to survive global climate change / Orrin H. Pilkey, Charles O. Pilkey, Linda P. Pilkey-Jarvis, Norma J. Longo, Keith C. Pilkey, Fred B. Dodson, Hannah L. Hayes.
Description: Durham : Duke University Press, 2024. | Includes bibliographical references and index.
Identifiers: LCCN 2023020141 (print)
LCCN 2023020142 (ebook)
ISBN 9781478025443 (paperback)
ISBN 9781478020660 (hardcover)
ISBN 9781478027577 (ebook)
Subjects: LCSH: Climatic changes. | Climate change mitigation. | Climatic changes—Effect of human beings on. | Global temperature changes. | Global warming. | BISAC: NATURE / Ecology | SCIENCE / Environmental Science (see also Chemistry / Environmental)
Classification: LCC QC903 .P557 2023 (print) | LCC QC903 (ebook) | DDC 363.7/06—dc23/eng/20231017
LC record available at https://lccn.loc.gov/2023020141
LC ebook record available at https://lccn.loc.gov/2023020142

Cover art: *Fire* courtesy Toa55/Shutterstock. *Sea Ice* courtesy Orrin Pilkey. *Pollution* courtesy Andreas Rentz / Getty Images. *Background* courtesy USDA (digitally remade by Charles Pilkey).

CONTENTS

PREFACE

The truth is that we have located our agriculture and built our communities and all associated infrastructure in places and to standards appropriate to what we thought was a relatively stable climate; climate conditions upon which we can no longer rely. What should be deeply alarming to all is that climate we have taken for granted for so long is, right before our very eyes, being replaced by a climate that, unless we act now, we may not survive. — Robert Sandford, Canadian water policy expert (quoted in Smith 2021)

Much has been published on the topic of global climate change, and we owe a debt of gratitude to these previous works. As we were writing this book in 2021–22, Nature taught us even more: wildfires in California, Oregon, Siberia, Europe, and China; heat waves in Europe, China, and North Africa; droughts in the southwestern United States, China, and Afghanistan; dwindling water levels in the Colorado, the Mississippi, the Yangtze, and the Rhine Rivers; major flooding in Yellowstone Park and Kentucky; Seoul, Korea; and Pakistan. Particularly memorable were:

- Death Valley (1.46 inches [3.7 cm] of rain, nearly a year's rainfall, in 3 hours!)
- Dallas (a summer's worth of rain in a day)
- Pakistan (millions displaced and much of the country covered by floodwaters)
- British Columbia (a deadly heat dome, followed by drought, wildfires, and floods)

All this plus a host of other climate-change-induced disasters are happening far sooner than scientists had anticipated.

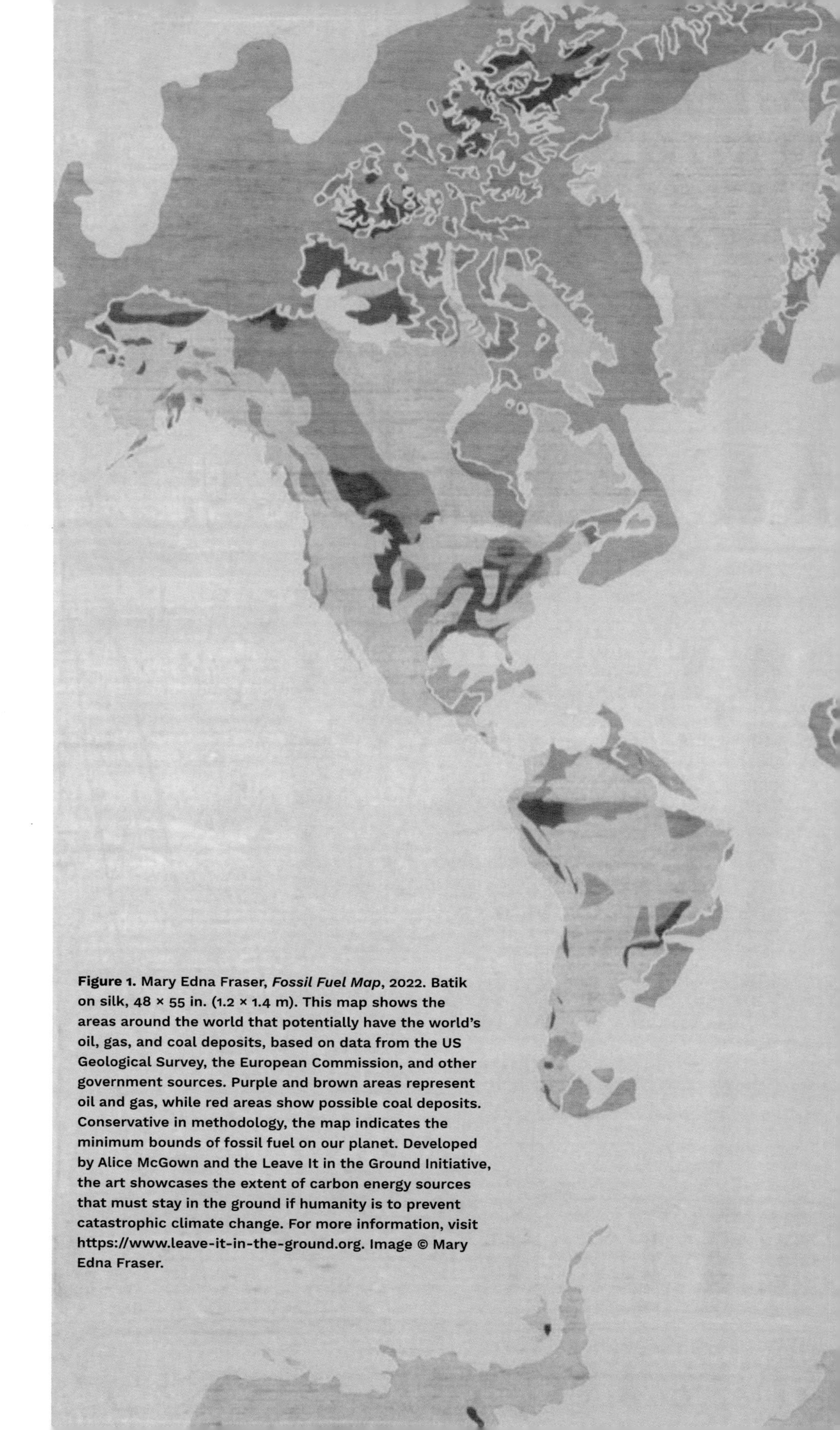

Figure 1. Mary Edna Fraser, *Fossil Fuel Map*, 2022. Batik on silk, 48 × 55 in. (1.2 × 1.4 m). This map shows the areas around the world that potentially have the world's oil, gas, and coal deposits, based on data from the US Geological Survey, the European Commission, and other government sources. Purple and brown areas represent oil and gas, while red areas show possible coal deposits. Conservative in methodology, the map indicates the minimum bounds of fossil fuel on our planet. Developed by Alice McGown and the Leave It in the Ground Initiative, the art showcases the extent of carbon energy sources that must stay in the ground if humanity is to prevent catastrophic climate change. For more information, visit https://www.leave-it-in-the-ground.org. Image © Mary Edna Fraser.

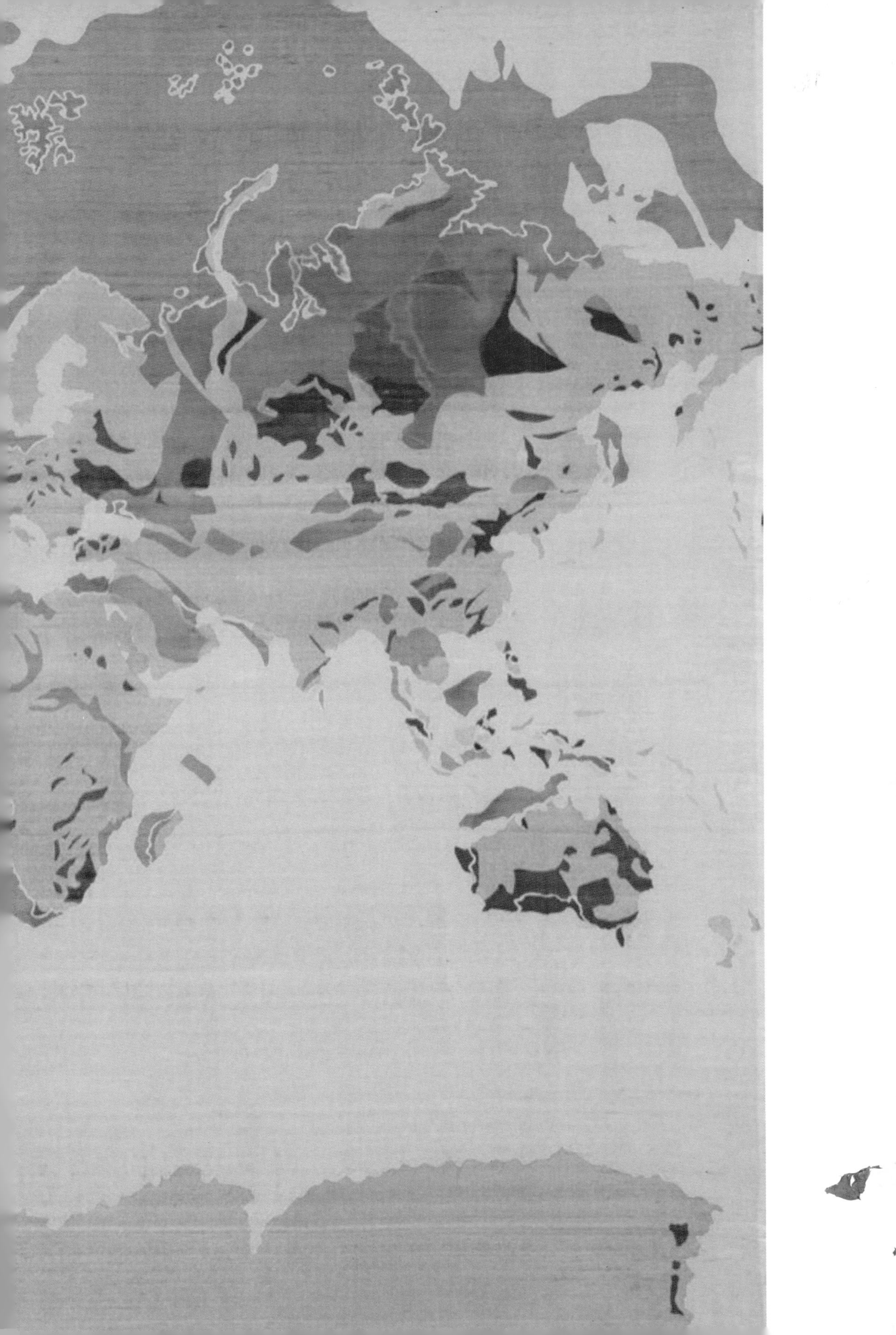

ACKNOWLEDGMENTS

This book was a team effort involving family and friends from a variety of backgrounds and skills including legal, scientific, environmental, educational, artistic, and political. The glue holding such a diverse group together was recognition of the huge threat that climate change holds for humans. The statement by Robert Sandford in the preface is one that we find deeply compelling.

In writing this volume, we received a lot of enthusiastic help from friends and others offering stories and information. More often than not, they had experienced a significant climate change event or knew someone who had. It seemed that everyone was interested in this topic.

We particularly wish to thank Gisela Fosado, editorial director at Duke University Press, whose enthusiasm for the book right from the start was an inspiration. Thanks also to assistant editor Alejandra Mejia and Lisa Lawley, our ever-patient project editor, who helped us navigate the technical problems of producing a publishable manuscript, and to Laura Sell, who got the word out.

INTRODUCTION

Greenhouse gas emissions keep growing. Global temperatures keep rising. And our planet is fast approaching tipping points that will make climate chaos irreversible. We are on a highway to climate hell with our foot still on the accelerator.
—António Guterres, secretary-general of the United Nations

Humanity has entered a nasty, brutish Hobbesian world where rising seas will drown the world's coastal communities, parts of the Earth will become too hot for humans to live, and millions of climate refugees will pour into the United States, Canada, and Europe, straining food, water, and energy supplies as well as human compassion. Droughts, floods, wildfires, powerful hurricanes, heat waves, food shortages, ocean acidification, and military conflicts over dwindling resources are among a host of evils released from the Pandora's box of industrial capitalism. Climate change has arrived. And humans will not be the only ones harmed.

As a species we are molded by natural selection to react to immediate physical danger, to run from a charging mammoth, or if need be to nock an arrow for defense. We are not mentally equipped to prepare for a slow-moving abstraction like climate change that unfolds over decades and centuries. It is no surprise that our minds become numb from unrelenting news of climatic disasters, paralyzed into inaction like deer caught in the headlights of climate dread, our fears amplified by an age of divided politics where one side refuses to acknowledge the findings of science simply because the other side does. But there is some good news and perhaps room for guarded optimism.

The mainstream media are finally starting to talk openly about climate change, something that would have been unthinkable just a few years ago. Moreover, we humans are intelligent, creative, and resourceful. We *will* do something about climate change. Why? Because we've been handed what is known in chess as a forced move, where a player must do something to save the king. Sooner or later the frequency and intensity of disasters triggered by a roasting planet will compel even the most hard-core, dyed-in-the-wool, obtuse climate change minimizer to admit that something is amiss, that a hitherto stable Nature has gone dangerously unhinged and we'd better do something about it or the game is lost. When that bright moment of clarity arrives, mammalian instincts to protect children at all costs will rise to the fore, overriding greed, political stalemate, and reluctance to change. Then at long last we'll roll up our collective sleeves and get to work.

And now we come to the crux of the matter. While we wait for society to reach that bright moment of clarity, what do we do in the interim? Anthropogenic climate change began with the rise of agriculture and animal husbandry before really taking off during the Industrial Revolution. We can't go back to a preindustrial state (nor would we want to). Our only options now for dealing with a warming world are to mitigate, adapt, or suffer. (See figure 2 for how these three options can sometimes overlap.)

Mitigation addresses the causes of climate change. It is a proactive attempt to remove carbon dioxide (CO_2) from the air either by using natural processes or through engineering. Examples of the former are planting trees, restoring ecosystems, and raising fewer children. Examples of the latter are developing carbon-free technologies and geoengineering.

Adaptation deals with the effects of climate change. It is a reactive approach that could be as simple as making a house fire resistant or as difficult as a family pulling up roots and moving to a region with no wildfires. In the case of coastal living, adaptation includes building seawalls, evacuating from hurricanes, or relocating entire cities inland.

Suffering is self-explanatory. We do little and so suffer much. For those with limited resources, the hardships of climate change will be hard indeed. Put simply, the poor will suffer the most. The rich will move to New Zealand.

The focus of this book is adaptation: what individuals, communities, and government can do to weather (no pun intended) the privations of an already deteriorating climate. If we are lucky, the apocalyptic events outlined above will unfold slowly over centuries instead of decades, giving us time to adjust. But the science suggests otherwise. The 2021 and 2022 reports of the Intergovernmental Panel on Climate Change (IPCC) conclude that the world's climate is changing faster than the models predicted and that the pace of change will accelerate in the near future. We are running out of time. While we wait for technological breakthroughs in climate mitigation and for the slow-turning wheels of bureaucratic governance to lumber toward some kind of clarity, we must fend for ourselves, must rethink how and where to live. To survive global change we must find ways to escape Nature.

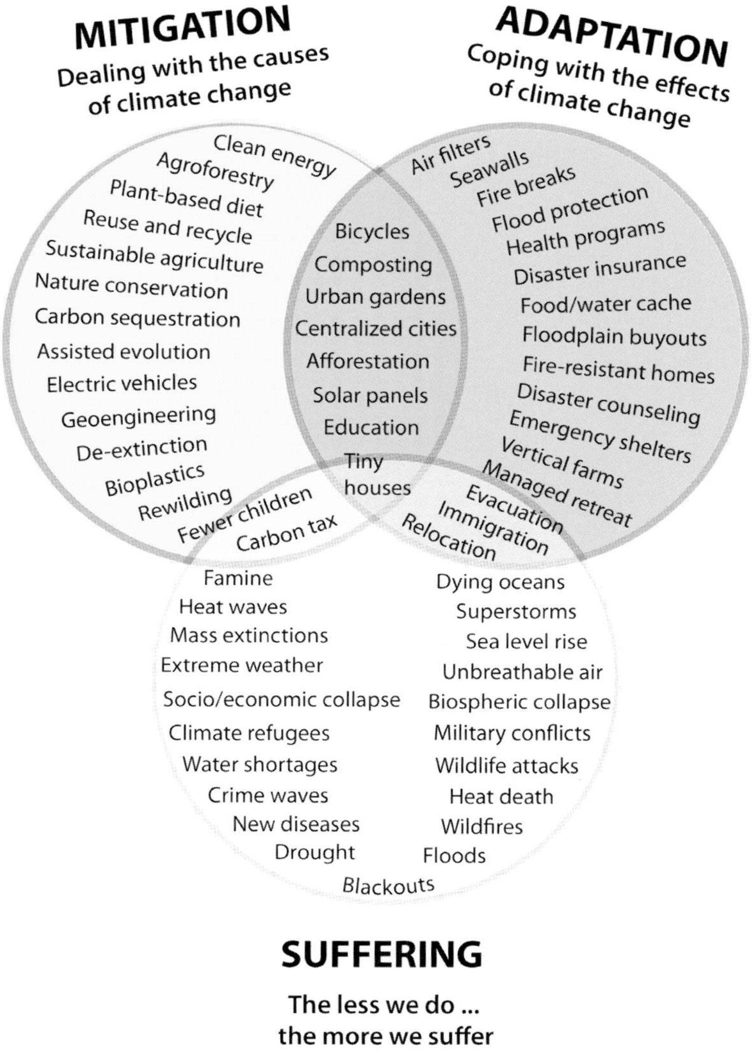

MITIGATION

Dealing with the causes of climate change

ADAPTATION

Coping with the effects of climate change

Clean energy
Agroforestry
Plant-based diet
Reuse and recycle
Sustainable agriculture
Nature conservation
Carbon sequestration
Assisted evolution
Electric vehicles
Geoengineering
De-extinction
Bioplastics
Rewilding

Bicycles
Composting
Urban gardens
Centralized cities
Afforestation
Solar panels
Education
Tiny houses

Air filters
Seawalls
Fire breaks
Flood protection
Health programs
Disaster insurance
Food/water cache
Floodplain buyouts
Fire-resistant homes
Disaster counseling
Emergency shelters
Vertical farms
Managed retreat

Fewer children
Carbon tax

Evacuation
Immigration
Relocation

Famine
Heat waves
Mass extinctions
Extreme weather
Socio/economic collapse
Climate refugees
Water shortages
Crime waves
New diseases
Drought

Dying oceans
Superstorms
Sea level rise
Unbreathable air
Biospheric collapse
Military conflicts
Wildlife attacks
Heat death
Wildfires
Floods

Blackouts

SUFFERING

The less we do ...
the more we suffer

Figure 2. The diagram illustrates the overlap between climate change adaptation, mitigation, and suffering. Drawing by Charles Pilkey.

The book is divided into five parts corresponding to the five elements of classical times: earth, air, fire, water, and space. Within each part are several chapters, or topics of discussion, which include (1) a description of a particular climate-related threat and (2) a what-to-do summary of ways to neutralize or adapt to that threat.

"Earth" recaps the geologic history of climate change and its impact on biodiversity, summarizes the findings of the UN IPCC reports, explains the significance of melting permafrost, and closes with a grim forecast of widespread crop failures, hunger, and malnutrition. "Air" is about greenhouse gases, changing weather patterns, heat waves, stronger hurricanes, and deteriorating air quality. "Fire" describes the impact of wildfires on human society and on the planet's ecosystems. Rising seas, dying oceans, floods, droughts, and declining water quality are some of the many cheerful topics covered in "Water."

The last part, "Space," talks about the migration of climate refugees (both human and nonhuman), health issues related to climate change, and how cities can adapt to a fast-changing planet. Its final chapter, "The Biosphere," concludes with an argument for the rights of Nature and a cogent reminder that we must maintain a healthy biosphere so that when we depart this world to colonize others, we will have a viable home planet to return to.

For those in a hurry, the book's main points are summarized in "The Heart of the Matter." "New Ideas" explores some of the more intriguing proposals for dealing with climate change. Because so many important climate-related events happened during the production of this book, we felt obliged to list them in "New Developments." "Bug-Out Bags" provides an inventory of essential items people should take when fleeing extreme weather events, followed by "To Learn More," a list of resources for those seeking a deeper understanding of climate science and biodiversity issues.

The way to climate stability will be hard. It will take all our ingenuity and determination. And then it will ask for more . . . and more. We, the children of the affluent West (and the affluent East), have been tasked with the near-impossible assignment of redesigning social, economic, technological, and even spiritual systems, sacrificing along the way our comfortable fossil fuel–based lifestyles so that future generations, people we will never meet or know, can lead happier lives free from the horrors of climate chaos. Some may be tempted to believe otherwise, but the strategies for adapting to climate change like those presented in this book are at best temporary measures, necessary tactics to buy time for policy makers to implement more permanent solutions. Adaptation is the beginning, but without mitigation it will not be enough. Either we curb our carbon emissions, or we suffer the consequences of societal and biospheric collapse.

EARTH

The Lessons of Geologic Time

Hence the same instant which killed the animals froze the country where they lived. This event was sudden, instantaneous and without any gradual development.—Georges Cuvier

The present is the key to the past.—Charles Lyell

The Evolution of the Earth

It's not hard to understand geologic time . . . it's impossible. What's a million years to a creature whose life spans less than a century? How are we to comprehend events that transcend generations of human lives or exceed the entirety of the time our species has walked the Earth? Gaze long into the abyss of geologic time and you will see nothing . . . only an abyss gazing back at you.

And so we construct rough metaphors to bridge the chasm of deep time and thereby make sense of the world. The geologic time scale is one such metaphor. Computer models, literature, and art are others. But a metaphor is only a metaphor, and reality is more complex than even our best mental constructs. And sometimes we become so enamored of metaphors that we fail to see truth, even when it stands right before our eyes.

In the 1800s geologists advanced two competing theories to explain how the Earth changes over time. The first was catastrophism, the belief that the Earth's surface is shaped by sudden, brief, and violent events. The opposing view was uniformitarianism, the theory that Earth-shaping processes unfold gradually in incremental stages. The catastrophists cited floods and volcanic eruptions to back their views. The uniformitarians looked at erosion and sedimentation to back theirs.

Eventually uniformitarianism won the day, and catastrophism was tossed into the waste bin of useless ideas, or so it seemed at the time. In the age of Darwin, when science was trying to wean itself from scripture, catastrophism was considered too "biblical" for scientists to take seriously. Thus, when paleontologists first noticed the dearth of fossils in the upper Permian strata, they failed to conclude the obvious, that a major and rapid extinction event had occurred. Likewise, the lack of fossils at the end of the Cretaceous failed to impress. Geologists interpreted that void to signify a gradual replacement of dinosaurs by other species. In those days, the notion of abrupt extinctions on a global level was beyond the ken of scientific dogma (see figure 3).

Everything changed in Copernican fashion in the 1970s when the physicist Louis Alvarez and his geologist son Walter Alvarez investigated a layer of iridium-rich clay in Europe. Iridium, a rare element on the Earth's surface, is far more common in space rocks like meteorites and asteroids. The iridium layer was 66 million years old, coincidentally the same age as the End Cretaceous extinction (aka Cretaceous-Tertiary or K-T extinction).

In 1980 the Alvarez team published a paper declaring that a killer asteroid had knocked off the dinosaurs. The proposal was initially greeted with howls of ridicule, but it gradually gained acceptance (as often happens with the advancement of science when new evidence backs up a new hypothesis). Scientists found the same iridium layer all over the world. And when a 66-million-year-old impact crater was discovered off the Yucatán coast, paleontologists were persuaded that something wicked from space had in fact this way come. The Alvarez team was vindicated.

But there's more to the story. The asteroid vaporized all life within a 620-mile (1,000-km) radius of the impact, unleashed massive tsunamis, and discharged so much dust into the atmosphere that the ensuing nuclear winter killed plants on every continent. When the plants died, the herbivores did the same, as did their predators. The blast was so powerful it shook the planet like a rag doll and triggered volcanism already underway in India, which in turn released enough carbon dioxide (CO_2) to precipitate a runaway greenhouse effect.

Global cooling followed by warming was too much for most life. Thousands of species vanished instantly with the blast. Millions more suffered a twilight of long decline as food sources dwindled and living conditions deteriorated. The world ended with a bang followed by a slow whimper.

The catastrophists had been right all along . . . or at least partly right. Now geologists recognize that the evolution of the Earth is a steady, gradual

Figure 3. The Geologic Time Spiral. Image from the US Geological Survey (USGS). Spiral designed by Joseph Graham, William Newman, and John Stacy (2008); digital preparation by Will Stettner. General Interest Publication 58.

unfolding punctuated by periodic catastrophes—asteroid strikes, volcanic eruptions, and mass extinctions.

Climate Change and Extinction

No life lives forever. Plants and animals evolve into new expressions or go extinct. Of the dozens of extinction events our planet has endured over the past 500 million years, five qualify as mass extinctions where 70% or more of all species vanished in a geologically short period of time. Scientists are still

Table 1. Mass Extinctions

Extinction event	Date	% of species made extinct	Examples of species made extinct	Causes
End Creta-ceous	66 mya	75%	Non-avian dinosaurs, ammonites, mosasaurs, plesiosaurs.	Global cooling from asteroid strike followed by global warming from high CO_2 levels due to increased volcanism.
End Triassic	201 mya	76%	All coral reefs went extinct. Conodonts, capitosaurs, and phytosaurs.	Global warming, ocean acidification, and ocean anoxia caused by basalt flows from breakup of supercontinent Pangea.
End Permian	252 mya	90%	Basically everything, including trilobites. More than 80% of marine life.	Runaway greenhouse effect caused by Siberian lava flows and methane emissions. Extreme heat, acid seas, drought, acid rain, anoxia.
Late Devonian	365 mya	75%	Graptolites. Biggest extinction of vertebrates, including armored fish and large sharks.	Global warming followed by an ice age (global cooling).
End Ordo-vician	440 mya	85%	Most marine invertebrates, including 100 marine families.	Global cooling (ice age) and subsequent lower sea level and anoxia.

Note: mya = million years ago.

working out the reasons behind those extinctions, but one thing at least is clear: *all five mass extinctions were caused by climate change* (and probably all of the lesser ones were as well). This fact alone should impart a sense of urgency as we try to reduce greenhouse gas emissions (see table 1).

The Sixth Extinction

It began 50,000-plus years ago after our ancestors migrated in large numbers out of Africa, colonizing lands that had never known modern humans. The result was mass extinctions everywhere, initially from overhunting but later from deforestation, habitat loss, and the introduction of invasive species. The First Americans, the Aborigines, and other groups we now call Indigenous eliminated 70%–80% of the megafauna from their respective continents. The pattern continued into historic times with the Polynesians, who in a few centuries of island hopping managed to wipe out 10% of the planet's bird species. The extinctions they started continue today in places like Hawaii and French Polynesia.

Currently around 8 billion people occupy nearly every habitable space on the planet. Cities and farms have engulfed entire ecosystems. The trickle of extinctions following our African exodus has exploded into a torrent, 100 to 1,000 times the normal background extinction rate. Human beings and their livestock now comprise a staggering 96% of the total biomass of mammals—more than all the whales, elephants, antelopes, giraffes, seals, rats, bears, kangaroos, and the rest combined.

Many believe we are on the cusp of a new mass extinction, the Sixth Extinction, similar in scale to the previous five mass extinctions. Others disagree, claiming the current wave of extinctions is not as severe as, say, during the end of the Cretaceous. But these are mostly linguistic arguments about what constitutes a mass extinction, and the clamor of scientific debate masks a most cogent lesson of geologic time: every major extinction event in the past reached a tipping point where the entire biosphere collapsed. Irrevocably. All the predator/prey relations, the symbionts, primary producers... everything died.

We are not yet at the tipping point of biospheric collapse. Most likely we have several decades, probably centuries, to get our act together before the world cascades into darkness. But the truth is that no one really knows. Perhaps the death of the world's coral reefs later this century will be the tipping point, at least for marine ecosystems. Or maybe the sudden release of methane from permafrost? Or the burning rainforests? Or... do we really want to wait and find out?

The United Nations (UN) reports that a million species are in danger of going extinct in the coming decades (IPCC 2021a). Droughts, wildfires, heat waves, and other agents of climate change will exacerbate the stress on wildlife. Life will react as it always has in the geologic past to the stresses of a warming world. Some species will move to cooler places. Some will adapt through speciation. And some will wither and die.

What Song the Parakeet Sang

The Carolina parakeet (*Conuropsis carolinensis*) once numbered in the millions, gracing the forests and canebrakes of the eastern United States (see figure 4). But settlers destroyed its habitat to make farms and ranches, leaving the bird nothing to eat, except of course farmers' crops. So the farmers began killing the parakeets indiscriminately. They quickly learned if they wounded a single bird, its cries of pain would draw more parakeets to the area, an unfortunate flocking response that allowed people to kill dozens

Figure 4. Carolina parakeet. Drawing by Charles Pilkey.

in a single day. By the early 20th century, the Carolina parakeet had gone extinct in the wild. In 1918 the last remaining specimen died alone in the Cincinnati Zoo, ironically in the same cage where the last passenger pigeon had died four years before.

It's easy when studying a subject as complex as biodiversity to get ensnared by the numbers game—the numbers of species going extinct every day, the percentage of habitat loss, and so forth. We forget that the extinction of a bird like the Carolina parakeet represents the loss of a sentient being, the death of a fellow traveler drifting on the stream of geologic time . . . the death of beauty. We'll never know what song the Carolina parakeet sang . . . and the world is a lesser place with its passing.

What to Do

- **Preserve Nature.** One of the subthemes of this book is that climate change not only afflicts humanity but also impacts Nature. We must recognize the importance of biodiversity and the need to prevent species extinctions. Our physical survival and the health of the planet's biosphere are one and the same.
- **Educate Leaders.** One reason the United States is failing to adequately deal with climate change is that our leaders are, more often than not, science illiterate. One cannot make rational, informed political decisions about anything from a position of ignorance. Environmental science (including the study of ecology and the evolution of life) should be taught at all levels of public education. Is it too much to ask that those running for public office be able to demonstrate some basic level of science literacy?

The 2021 United Nations Climate Report

The UN wasn't created to take mankind into paradise, but rather, to save humanity from hell.—Dag Hammarskjöld, second secretary-general of the United Nations

The United Nations (UN), formed at the end of World War II in a time of desperate hope for a lasting world peace, has expanded its role to include concerns about the future of our planet's climate. The UN Intergovernmental Panel on Climate Change (IPCC), created in 1988, has successfully consulted with some of the world's most talented and least politically affiliated scientists to map the most likely future of climate change.

The IPCC's August 2021 report, their Sixth Assessment Report (AR6), has proven to be a bombshell—a nail in the coffin of the concept that the massive global climate change that we face isn't all that bad. This UN report, plus the occurrence of so many extreme weather disasters in recent years around the world (e.g., the California and Canada wildfires, heat waves, catastrophic rainfalls, the rising sea level, droughts, and the increasing numbers of climate refugees), have led to a much greater appreciation of the existential threat that climate change presents to humanity.

Major Points in the Report

- **It's already here.** Human-induced climate change is already causing many extreme climate events in every region across the globe.
- **The Paris agreement won't save us.** We face a "Catastrophic Pathway" even if all the countries follow their promises made in the 2015 Paris Climate Agreement to reduce greenhouse gas emissions.
- **We are the problem.** The evidence shows that industrialized nations, the United States and China, in particular, are producing greenhouse gases that cause global warming.
- **As the climate gets warmer, disasters increase.** Climate extremes are larger in area, more frequent, and more intense with every additional increment of global warming. Thus, it is crucial that we eliminate or at least reduce the emissions of greenhouse gases to preserve a livable planet.
- **We should prepare even if we reduce CO_2 emissions.** We are locked into significant climate change even if we significantly reduce greenhouse gas emissions, since many changes due to greenhouse gases will be irreversible for centuries to millennia. These include changes in the oceans, ice sheets, and global sea level. Thus, it would be wise indeed to prepare for climate-related disasters on all fronts: personal, familial, community, national, and international levels.
- **We could be wrong; it could be worse.** The possibility of global disasters such as ice-sheet collapse (the Thwaites Glacier in Antarctica) and abrupt ocean circulation changes (the halting of the Gulf Stream by freshwater flow from Greenland's melting ice) cannot be ruled out and should be part of any risk assessment.
- **Things will happen fast for a long, long time.** The rate of recent changes in the climate has increased (satellite imagery has indicated that the rate of sea level rise has doubled in two decades). Climate changes happening now will continue unfolding over many centuries to many thousands of years.
- **Climate change is a "Code Red for humanity."** This assessment is according to UN Secretary-General António Guterres.

What's to Come

CARBON DIOXIDE LEVELS

Concentrations of carbon dioxide (CO_2) in the atmosphere are higher now than at any time over the past 2 million years. Other greenhouse gases such as methane (CH_4) and nitrous oxide (N_2O) now exceed any known concentration over the past 800,000 years. In the short term (20 years),

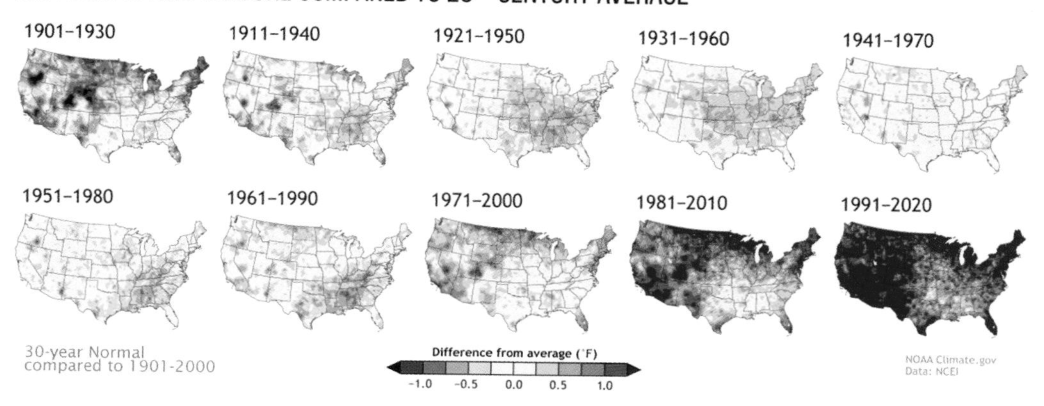

Figure 5. Temperature changes: The red tones represent regions where average annual temperatures for each three-decade period are higher than the 20th-century average. Blue indicates regions where the average annual temperatures are lower than the 20th-century average. Image from NOAA.

methane has about 80 times the warming power of carbon dioxide, which is why the report urges cutting both methane and carbon dioxide emissions. Permafrost covering 25% of the Northern Hemisphere is releasing long-trapped greenhouse gases at an increasing rate. Assuming that we continue to emit large quantities of CO_2, the carbon sinks both in the ocean and on land will be overwhelmed and hence less effective in reducing climate change.

RISING ATMOSPHERIC TEMPERATURES

Each of the past four decades (1980–2020) has been successively warmer than any decades since 1850. The atmosphere is warming faster now than at any time during the past 2,000 years. Atmospheric temperatures will increase more than 3.6°F (2°C) this century. Continuing warming will increase the frequency and intensity of heat waves, hurricanes, marine heat waves, heavy rainfalls, agricultural and ecological droughts in some regions, as well as reductions in Arctic sea ice, snow cover, and permafrost (see figure 5).

ICE MELT

The world's ice (mountain glaciers, sea ice, ice sheets) is melting. The extent of late-summer Arctic sea ice is smaller than at any time in the past 1,000 years. Sea ice protects Arctic shorelines, and its absence will increase shoreline erosion during storms. The loss of sea ice's reflectivity of the sun's radiation adds to the warming of the Arctic, which already is heating up at four times the rate of warming at lower latitudes. Almost all the world's glaciers have been retreating since the 1950s. The highest rate of glacier ice loss is occurring in the southern Andes, central Europe, Alaska, and Iceland. The loss of glaciers makes a powerful case for the need to reduce emissions as soon as possible, as glaciers are important sources of domestic water supply in much of the world.

SEA LEVEL RISE

The rate of sea level rise is the highest it has been in the past 3,000 years. Between 1901 and 2018, the rate of global mean sea level rise was 1.3 mm (0.05 inch) per year but between 2006 and 2018 the rate increased to 3.6 mm (0.14 inch) per year. Sea level is set to rise for centuries to millennia, due to continuing ocean water expansion when heated, plus the melting of ice sheets and glaciers. According to the UN IPCC report (2021a), sea level rise could be between 5 feet (1.5 m) and an extreme rise of 8.2 feet (2.5 m) by the end of this century.

OCEAN ACIDIFICATION

Carbon dioxide dissolved in seawater makes the ocean more acidic and threatens sea life. Damage is particularly visible on coral reefs. Acidification threatens seafood availability and could be a factor in creating famines.

HEAT WAVES

Heat waves on land have become more frequent and more intense since the 1950s. Simultaneously, cold waves are becoming less frequent and less severe. Heat waves that used to happen once every 50 years are now happening every decade. Marine heat waves (the ocean equivalent of a heat wave on land) have nearly doubled since the 1980s.

PRECIPITATION

A warmer atmosphere holds more water and has led to heavier precipitation events since the 1950s (see figure 6). Extreme rain/snow events will continue to increase. Overall, the Earth will get wetter, most likely in parts of Africa, India, and the northeastern United States. However, some areas will probably get drier, such as the southwestern United States, Central America, the Amazon, southern Africa, the Mediterranean, and the Middle East. These regions will suffer major wildfires as well as water shortages.

DROUGHTS

Warming leads to increased evaporation from soils and vegetation and contributes to droughts. Just as heavy precipitation events are expected to increase, so will droughts, a major challenge to food production and

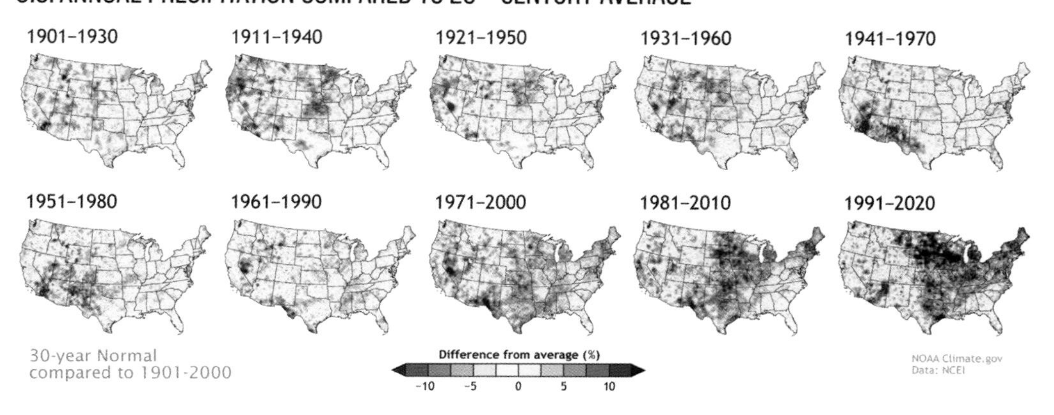

Figure 6. Precipitation changes: this figure illustrates how precipitation has increased relative to the beginning of the 20th century for three-decade periods. In the mid-20th century, rainfall was reduced, but by the end of the century and beginning of the 21st century, rainfall strongly increased. Image from NOAA.

drinking-water supplies in coming decades. The frequency and intensity of droughts will increase in some regions, including the Mediterranean, West and Northeast Asia, much of South America, and Africa. The frequency of dust storms has increased, with negative impacts on human health in dry regions such as the Arabian Peninsula and Central Asia. In sub-Saharan Africa, Central Asia, and Australia, warmer temperatures combined with the intense grazing of grasslands by sheep, goats, and cattle are leading to desertification.

FOOD SECURITY

Food availability has been affected by climate changes in dry lands in Africa as well as in high mountainous regions of Asia and South America. The tropical and subtropical regions are projected to be the most vulnerable to crop-yield declines. Warming temperatures, rainfall variations, and the frequency of extreme weather events have already reduced food production and animal growth rates. Also, the responses of agricultural pests and diseases are changing along with the global climate.

Changes in food consumption patterns have contributed to about 2 billion adults (and many children) being overweight or obese, while an estimated 821 million people are undernourished. In addition, urban expansion is expected to continue taking up cropland, leading to greater losses in food production.

TROPICAL CYCLONES (HURRICANES)

Over the past four decades, hurricanes have become more intense, with increased proportions of Category 3–5 storms, and with heavier rainfall. Also, hurricanes are larger, are moving more slowly, and change from tropical storms to hurricanes more rapidly, thus making the path of these storms more difficult to predict.

COP26 and COP27

The 26th session of the Conference of the Parties (known as COP26) to the UN Framework Convention on Climate Change (UNFCCC) was held three months after the release of the UN's IPCC 2021 climate change report. However, this time the event was unabashedly about politics as 120 world leaders along with hundreds of representatives of governments, environmental organizations, and businesses met in Glasgow, Scotland, in early November 2021. The urgency of climate change was apparent throughout the COP26 meeting and described in a speech by then-UK prime minister Boris Johnson as "a detonation that will end human life as we know it" (Johnson 2021).

The meeting was planned to address efforts to fulfill the goal of the 2015 Paris Agreement, that is, to stop the planet from warming more than 1.5°C (2.7°F) above the globe's preindustrial level. This would require halving greenhouse gas emissions by 2030 and reaching zero net emissions by 2050, monumental accomplishments indeed. The Glasgow Climate Pact is the result of the 14 days of meetings. Two major commitments are to reduce methane emissions (mostly from the oil and gas industry, livestock, and landfills) and to stop deforestation. A number of methane polluters, among them China, Russia, Australia, and India, refused to join the methane agreement. Moreover, the leaders of Russia and China (the Chinese sent a written statement) did not attend COP26, which doesn't bode well for the success of the conference. Other nonattendees were Brazil, Iran, Japan, Mexico, South Africa, and Turkey.

The COP27, held in Egypt in the fall of 2022, was a continuation of discussions started at COP26 (Alayza et al. 2022). Putin and Xi again chose not to attend. This is unfortunate as China is the world's biggest emitter of CO_2 while Russia is the world's fourth largest emitter, behind the United States (second largest) and India (third).

Perhaps the most important breakthrough at COP27 was the agreement to establish funding to help those poor countries most vulnerable to climate change. The UN also proposed a $3.1 billion plan to offer early-warning systems to every nation on Earth so people everywhere can prepare for imminent extreme weather events.

Famine

Food is national security. Food is economy. It is employment, energy, history. Food is everything.—Chef José Andrés, founder of World Central Kitchen

The Causes of Famine

It's an old story... the struggle of life to fend off hunger. If history were a symphonic poem, then famine would be its leitmotif. According to the United Nations (UN) World Food Programme (WFP 2021), famine is underway when malnutrition is widespread, and people are dying from starvation. The traditional causes of mass hunger are legion... war, overpopulation, politics, poverty, crop failures, natural disasters, soil degradation, and a host of other factors stemming from human cruelty, Nature, or bad luck. Now we must add climate change to that list.

The Great Nutrient Collapse

Climate change will bring about famine via several mechanisms, the most obvious being crop failures and livestock loss due to floods, wildfires, droughts, and higher temperatures that favor insect infestations. To an extent this will engender the kind of episodic food shortages that humans have always struggled to surmount. But global warming will intensify food insecurity, in part because higher temperatures (with some regional variations) reduce crop yields and also because warming, acidifying oceans will degrade marine ecosystems, shrinking the food supply for hundreds of millions of people worldwide. All these events will unfold as our population climbs to perhaps 11 billion by century's end. More people. Less food... more hunger.

But there is another insidious connection between climate change and food security. Globally about two-thirds of human calories come from corn, soybeans, and grains. As carbon dioxide (CO_2) levels rise, the nutrients in plants decline. More atmospheric CO_2 makes plants bigger but at the same time increases levels of carbohydrates while depleting levels of protein, vitamins, zinc, iron, calcium, and other nutrients, turning healthy fruits and vegetables into junk food. This was first made known in 2004 through the seminal research of Irakli Loladze, a mathematician with a penchant for biology. In *Politico* he describes plant nutrient losses from rising CO_2 levels as "the great nutrient collapse" (Evich 2017).

One study projects a 19.5% reduction in plant proteins by 2050 and a 14% decline in iron and zinc. Perhaps as many as 150 million people in low-income countries will experience protein deficiency from nutrient collapse by midcentury. Hundreds of millions more will likely suffer from vitamin and mineral deficiency. The same phenomenon will impair the health of herbivores, including our livestock. It will also hurt wildlife. Scientists speculate that the worldwide decline in bee numbers is partly due to reduced protein levels in goldenrod, a late-summer food favored by bees.

A Plague of Pests

Made famous by the biblical plagues in Exodus, locusts and famines share a long history. Locusts are actually grasshoppers that congregate in search of food after undergoing rapid population bursts. According to some estimates, a large swarm can eat the same volume of food in one day as 35,000 people.

In 1874 a single swarm of Rocky Mountain locusts (*Melanoplus spretus*) eclipsed the skies from Canada to Texas, an area 1,800 miles (about 2,900 km) long and over 100 miles (160 km) wide. The estimated population of 12.5 trillion individuals is the planet's largest known concentration of animals. They ate everything—trees, wood, leather, wool on living sheep, even the clothes on people's backs (Ryckman 1999).

For more than a hundred years, the Rocky Mountain locusts laid waste to farmers' crops until unexpectedly going extinct in the early 20th century (Ryckman 1999). Biologists still puzzle over how a species so numerous could suddenly disappear. The most likely explanation is that plowing disturbed the

Figure 7. A swarm of locusts in Kenya in 2020. Photo © FAO / Sven Torfinn.

soil where the locusts laid their eggs. More worrying are the issues implicit when a species numbering in the trillions goes extinct in mere decades. What does that mean for less numerous organisms like monarch butterflies?

It is unlikely that we will again see massive swarms blanketing Earth's skies like those in 19th-century America. But in many regions locusts are still a formidable force of Nature, shattering local ecosystems, generating mass hunger, and darkening the dreams of farmers.

In 2020 extreme heat interspersed with heavy rains in East Africa created ideal breeding conditions for the desert locust (*Schistocerca gregaria*). Hordes numbering in the billions swarmed north and east from the Horn of Africa and ate their way across the Middle East all the way to India, making harder still the hard lives of farmers already burdened by poverty, unsafe drinking water, and the COVID-19 pandemic. In recent years similar swarms have menaced Australia, Russia, Argentina, Sardinia, southern China, South Africa, Madagascar, and parts of West Africa. According to the UN Environment Programme, a hotter, wetter climate will lead to more frequent plagues of locusts (UNEP 2020) (see figure 7).

Grasshoppers aren't the only biological burden farmers face. Hundreds of other insect species ravage crops worldwide. Some, like the corn rootworm, eat

Figure 8. A swarm of locusts are all feasting on a single plant. Chances are they will make short work of it. Photo from Kenya's Samburu County, near Wamba, 2020. Photo © FAO / Sven Torfinn.

specific plants. Others, like the invasive Japanese beetle (the bane of American gardeners), target anything from grass and grapes to fruit trees. Warming temperatures will fuel insect hunger by increasing their metabolism. Warmer winters will produce fewer cold snaps that might otherwise kill off insect larvae.

Today insects destroy 5%–20% of the world's major grains. But that number is expected to increase 10%–25% per degree Celsius (1.8 degrees Fahrenheit) of warming (mostly in temperate regions), further destabilizing our food security. And crop pests are creeping toward the poles around 2 miles (3 km) each year as the climate continues to warm (see figure 8).

Recent Famines

The largest famine in modern history, China's Great Leap Forward (1959–61), resulted in from 24 to perhaps 50 million deaths when Mao Zedong tried to transform China from an agricultural economy to an industrial one overnight. Currently several countries are on the brink of famine: Nigeria

(2 million), Somalia (3 million), Sudan (5 million), and Yemen (14 million). Some would include Afghanistan, Ethiopia, the Congo, South Sudan, North Korea, and Venezuela on the list.

We've already seen two recent famines sparked by climate change. The first was when extreme drought in Syria helped precipitate civil war, which in turn led to mass starvation during the siege of Aleppo. The second is happening at the time of this writing in Madagascar, the result of the country's worst drought in 40 years. Over 1 million people there are enduring catastrophic levels of hunger. Many are eating insects and cactus leaves to survive.

The terrible suffering in Syria, Madagascar and elsewhere is a prelude of worse things to come should we fail to curb carbon emissions. Extreme weather, nutritional deficits, and insect scourges will trigger food shortages and exacerbate existing political tensions. This century will likely be one of great hunger.

What to Do

- **Find ways to offset food scarcity.** The face of the future will be the face of hunger. Conflict and extreme weather will bring failed harvests, higher food prices, disrupted food supply chains, food shortages, and famine, especially in poor countries. Wealthy nations have a degree of economic cushion, but lower-income families in developed countries like the United States will still face periodic food scarcity. When times get hard, you can alleviate some of your food insecurity by growing a vegetable garden, raising chickens, or fishing in nearby rivers, ponds, lakes, and oceans. You can also find sustenance in wild fruits and vegetables (mainly weeds) such as dandelions, blackberries, cacti, wild grapes, persimmons, plantain (*Plantago* spp.), chicory, cattails, and other wild edibles (see figure 9). Harvesting these plants should be seen as a temporary means to keep a family going in an emergency until the normal food supply chain is reestablished. Purchase a field guide to learn what is edible and what isn't.
- **Conduct more research on nutrients.** The nutrient collapse is a recently discovered phenomenon. More research needs to be done about declining nutrients in crops and how that decline might affect pollinators and ecosystems. Should we take vitamin supplements? Should vegetarians be worried? How will nutrient collapse affect the health of the already nutrient-stressed poor? Will it impact biodiversity?

Figure 9. Prickly pear cactus. Drawing by Charles Pilkey. Prickly pear (*Opuntia* spp.), a cactus native to the Americas, grows in much of Canada, the United States, Mexico, the Caribbean, and South America. The fruits are sweet and juicy. Use gloves when harvesting. Use a knife to cut away the skin, which harbors tiny hairlike spines. Enjoy.

- **Have some backup food sources.** For the citizens of wealthy nations, famine is usually a distant memory. But one of the lessons of the COVID-19 pandemic is that supply chains can unexpectedly be interrupted, leading to skyrocketing food prices and empty shelves in the supermarket. There is no food security in an age of changing climate. People should stockpile at least a 2-week supply of food and water; if possible, cultivate a vegetable garden, and learn what wild edibles are locally available.
- **Develop new ways to control locusts.** The traditional way to deal with locusts is with insecticides. But insecticides kill all insects, including those that sustain birds and those that pollinate our food. Scientists need to devise population-control tools specific to locusts.
- **Use solar dryers.** Solar dryers are cheap, energy-efficient ways that farms in developing countries can become more climate resilient and less vulnerable to insect predation (see figure 10). Harvests can be quickly dried, stored, and then sold in the offseason, potentially offsetting economic losses due to insects, floods, and mold. Some small-scale farmers in America are using solar dryers for drying fruits, vegetables, spices, and medicinal plants.

Figure 10. Sun-powered dryers like this one in Bolivia are used by impoverished farmers. The solar dryers preserve and store food to avoid insects and spoilage, both for home use and for taking to market. Photo by Natalie Pereyra, GIZ EnDev Bolivia.

- **Help support relief efforts.** Perhaps the subject is close enough to your heart that you may wish to donate to relief organizations like the Red Cross or start your own fundraising organization. Go to the Network for Good website (https://www.networkforgood.com/) to get coaching for starting an online nonprofit charity.

Permafrost

The ice caps are melting,
the tide is rushing in.
All the world is drowning,
to wash away the sin.
—Bill Dorsey, "The Other Side"

The Changing Arctic

In Inuit mythology the Qalupalik (figure 11) is a kind of mermaid, a slimy, green-scaled monster lurking beneath thin ice that kidnaps misbehaving children and pulls them into the cold sea. It is said the monster makes a humming sound to lure kids to the water's edge. For untold generations Inuit parents have warned their children to stay off the ice lest the Qalupalik wrap its clawed hands around them and steal them away. But such myths are more than just entertaining stories. They are stratagems crafted to discipline the young and teach them how to survive in a harsh and dangerous land.

Today the Arctic is changing fast. The ice is melting. Cruise ships ply the now ice-free summer waters of the Northwest Passage. Even the ground itself is changing and can no longer be trusted. The old ways are going. The old stories are losing their meaning (see figure 12).

Figure 11. Qalupalik. Drawing by Charles Pilkey.

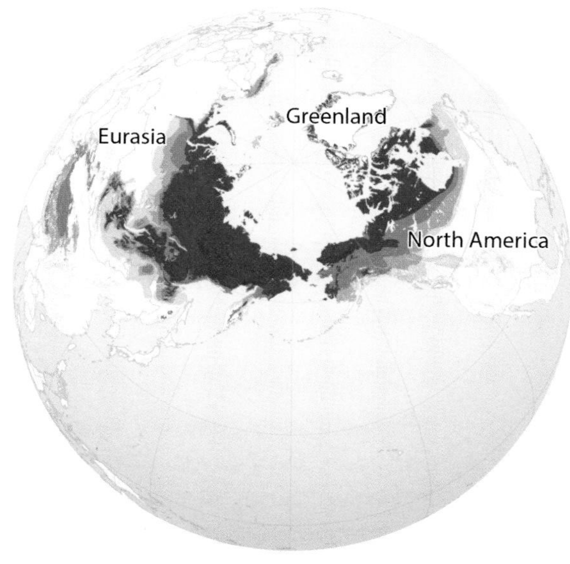

Figure 12. Around 25% of the Northern Hemisphere is covered by permafrost, which is ground that remains frozen for at least 2 years. Permafrost ranges from a few inches to miles in thickness. As the world's permafrost continues to melt, large quantities of methane and CO_2 are released into the atmosphere. Image from NASA.

Symphony in C Major

Permafrost is any ground that remains frozen for 2 years or more. Layers of permafrost can range from a few feet to more than 3,000 feet (1,000 m) in thickness and are often tens of thousands of years old. Found at high mountainous elevations and more commonly near the North and South Poles, permafrost covers 25% of the land in the Northern Hemisphere (more if you count permafrost under the seafloor), including large swaths of Siberia, Scandinavia, Greenland, Alaska, and Canada.

But now climate change is shattering the millennia-long stability of Arctic ecosystems, deforming the land, melting ice sheets, unleashing the fury of rising seas, and thawing permafrost. This last item has sober implications for Earth's climate since massive amounts of carbon dioxide (CO_2) and methane (CH_4) are locked in frozen ground. According to one estimate, permafrost soils globally contain around 1.5 trillion tons (1.36 trillion metric tons) of CO_2, roughly twice what is currently in Earth's atmosphere.

Scientists usually measure atmospheric CO_2 levels as parts per million (ppm). Before the Industrial Revolution, CO_2 levels were about 280 ppm. In other words, every one million molecules in air included 280 molecules of CO_2. Today the concentration of CO_2 is over 400 ppm (424 ppm in May 2023). That number varies seasonally but has been steadily rising every year since the 18th century. What keeps climate scientists up late at night is the possibility of a massive rush of CO_2 into the atmosphere after a sudden, precipitous melting of permafrost.

As permafrost melts, CO_2 and methane are released, warming the atmosphere, which in turn melts more permafrost and vents yet more carbon, thus triggering a classic positive feedback loop, the same feedback loop our planet has endured many times in the geologic past (see "The Lessons of Geologic Time" in "Earth"). In the ensemble of geochemical players orchestrating climate change, carbon is the principal conductor. It is the master of the situation.

The End of the Land

For decades Siberians have chanced upon mysterious craters, some 100 or more feet (30 m) deep, scattered across the tundra. In 2017 reindeer herders reported hearing an explosion, followed by flames and smoke spewing from one such crater on the Yamal Peninsula ("Yamal" means "end of the land"), roughly 1,400 miles (2,200 km) northeast of Moscow. Scientists theorized that meteorites had formed the craters. Others reckoned missiles or mischievous aliens were to blame. But after high levels of methane were

Figure 13. A large crater on the Yamal Peninsula in northern Siberia, formed in 2014 by the explosion of the methane gas released by melting permafrost in a process known as *cryovolcanism*. Numerous such craters are found on the peninsula and on the Arctic seafloor; more are expected in the future as the atmosphere warms. Photo by Ruslan Amanzhurov (Creative Commons CC BY license).

detected on the crater floors, it became clear that a warming climate was the culprit (see figure 13).

Methane is formed from geological processes and from biological ones (anaerobic bacteria in rice fields, landfills, and the guts of ruminants). Nearly 90 times more potent a greenhouse gas than CO_2 over a 20-year span, methane in permafrost is trapped within the crystal structure of frozen water (ice), forming what are variously called methane hydrates or methane clathrates (or colloquially "fire ice"). Now the hydrates are melting on land and under the seafloor, releasing increasing amounts of methane proportionate to rising temperatures and intensifying heat waves.

To date, no one has actually witnessed the genesis of a methane crater from close quarters (which may be a good thing). Scientists believe the craters are caused by methane collecting near the top of naturally forming mounds called *pingos* (figure 14). A pingo is created when hydrostatic pressure exerted by water and ice gradually pushes tundra up into a conical mound. It's theorized that methane itself may supply the requisite pressure to form some mounds. In any case, at some point the mound collapses, releasing methane in a sudden burst that can fling chunks of tundra hundreds of feet in all directions, occasionally igniting in a fiery inferno. What happens when

Figure 14. The Ibyuk Pingo in Canada's Northwest Territory, 161 feet (49 m) high. Photo by Adam Jones (Creative Commons Attribution-ShareAlike 2.0 Generic license).

a methane mound explodes near a gas pipeline or in a town is anyone's guess. Possibly we will find out one day soon.

Arctic Meltdown

Perhaps nowhere else is the global nature of climate change better demonstrated than in the Arctic. Melting permafrost from Siberia to Alaska is disrupting roads; changing traditional fishing, herding, and hunting patterns; and causing buildings to tilt drunkenly and in some places to sink and even collapse (figure 15). Some communities have turned into swamps where there is no place to bury the dead. Wildfires are spreading over once permanently frozen grounds in Siberia, Canada, and Greenland. Warming temperatures are allowing animals to extend their ranges north into the Arctic (see "Nature on the Move" in "Space"), including grizzly bears, which are interbreeding with polar bears, producing a hybrid called by some the *pizzly bear*.

Warmer Arctic temperatures are also starting to revive long-dormant pathogens frozen in the carcasses of reindeer and other animals. This is what

Figure 15. A building collapsing and sinking in Fairbanks, Alaska, under the influence of melting permafrost. This is a widespread and much-feared phenomenon in Siberia, Alaska, and Canada. Photo © Vladimir Romanovsky, professor emeritus of the Geophysical Institute, University of Alaska Fairbanks.

happened in 2016 when a Siberian anthrax outbreak spread from a thawed reindeer carcass to infect a community of nomadic herders and more than 2,000 of their reindeer (see "Health" in "Space"). In 2015 French scientists were able to "reawaken" a 30,000-year-old virus frozen in the Siberian permafrost. While that particular virus is harmless to people, scientists speculate there may be viable "zombie pathogens" oozing out of the melting Arctic. Some, like anthrax, may be infectious to humans. And some may be bacteria and viruses that modern humans have never encountered.

In Alaska temperatures are rising faster than in any other US state. Previously stable slopes have become unhinged, transformed into slow-moving underground landslides known as *frozen debris lobes*. These slow-motion landslides composed of rocks, soil, and partially thawed permafrost can pull down trees, wreck buildings, and destroy highways and pipelines. Advancing up to 20 feet (6 m) a year, they are unstoppable. In 2018, a 4,000-foot (1,220 m) section of the Dalton Highway (the road that provides access to the Trans-Alaska Pipeline) had to be moved to avoid a debris flow. It will likely have to be moved again in a few years. Engineers are now trying to conjure methods to keep the Alaskan pipelines safe from frozen debris lobes.

The economic costs of adapting to climate change in the Arctic are staggering. This is especially the case in Russia, where nearly 65% of the country is in the permafrost zone. Road repair alone between 2020 and 2050 will cost an estimated $365 million. Because most Russian pipelines are underground, they are particularly vulnerable to land sinking from melting permafrost. If emissions remain unchanged, the cumulative costs of repairs to the natural gas pipelines could be $110 billion, roughly equivalent to one year of revenue from Russia's gas fields. At that price, is it possible some gas fields will be abandoned?

Shoreline Erosion

Melting sea ice is on the news a lot these days with regard to the plight of polar bears, which depend on the ice to hunt seals. It's not clear how polar bears will survive an ice-free Arctic. Human communities will be hard pressed as well. Native Alaskans who traditionally harpoon whales from boats launched from shore are having a hard time finding whales. The thinning ice pack is allowing the whales to migrate further offshore. With thinning ice and warming water, orcas (killer whales) are expanding their range into Arctic waters. Feasting on fish, seals, and whales, orcas are disrupting a food web that has sustained marine species and human communities for millennia.

Thinning sea ice has repercussions beyond affecting the lifestyles of Indigenous coastal communities. It will seriously impact their physical survival as well. At least 31 towns in northern Alaska are in danger of inundation from beach erosion made worse by melting sea ice and thawing permafrost on beaches (figure 16). Examples include Shishmaref (population 600) and Kivalina (population 400), two mostly native villages on barrier islands along the Chukchi Sea coast of Alaska. Before the advent of global warming, sea ice piled up next to the beach in early fall, forming a natural barrier that protected shorelines from the heavy surf and storm surge of November gales. But now the ice doesn't freeze over until December and January. Storm waves are eroding an unfrozen, unprotected beach, threatening to overrun northern Alaska's seaside communities.

Shoreline erosion accelerated by melting permafrost has already forced the community of Newtok to relocate. This Native Alaskan village along the Bering Sea is moving (with federal support) to more solid ground, building a new town called Mertarvik. But relocating an entire village in Alaska is not cheap due to the high cost of shipping in machinery and other supplies. To transplant a town the size of Newtok to safer ground will cost taxpayers around $100 million. And that's for a mere 400 people!

Figure 16. A winter photo of a house fallen onto the beach in Shishmaref, Alaska, along the Chukchi Sea. Sea ice that once protected the shoreline from autumn storm surf is disappearing. The once-frozen beach sand is thawing, permafrost on land is melting, and to top it off, the sea level is rising. No wonder Inupiat communities like Shishmaref are no longer protected from the encroaching sea. Photo by Orrin Pilkey.

Life for the nearly 5 million people who live on Arctic and subarctic permafrost will never be the same. Their world is changing forever. Many will have to move elsewhere. The ground itself is shifting beneath their feet. Shorelines are eroding. Tundra is discharging methane. And some say under the fast-thinning ice a strange humming sound can be heard . . . the Qalupalik is stirring.

What to Do

- **Conduct research.** More research is needed to better predict how, when, and where methane explosions might occur, and how to stop frozen debris flows.
- **Revise guidelines for new buildings.** To build safely in permafrost regions, one should (1) put structures on bedrock, (2) make sure ground is well drained so water can't accumulate and make the area soggy with spring melting, (3) place buildings on pilings above the ground so cold air circulates in the open space underneath to keep

the ground frozen, and (4) clear snow cover as it accumulates in winter. Snow insulates the ground. Once the snow blanket is gone, the soil will be cooled by the air and more likely stay frozen.

- **Stabilize existing structures.** House jacks can be used to temporarily stabilize structures on permafrost. A high-tech solution uses Thermosyphon tubes that keep the ground cold by transferring heat from the ground to the air.
- **Provide support for relocation.** More money needs to be allocated to state and federal agencies to finance the relocation of climate refugees such as those in northern Alaska.
- **Embrace new lifestyles.** The Arctic is changing so fast that hunters, trappers, fishermen, pastoralists, and others pursuing subsistence lifestyles will have to adapt accordingly. This could entail shifting the seasonal timing of food gathering, finding new food sources, or abandoning old lifestyle patterns.

AIR

Hurricanes

Shortly before the storm's arrival, strange weather had settled on the island. The day was intensely hot, the sky rimmed with a reddish-yellow light. There was, according to the Antigua Standard, an "ominous" stillness.—Erik Larson, *Isaac's Storm*

Hurricane Forecasts

On September 5, 1900, when a tropical storm passed over Cuba, the meteorologists there alerted the US Weather Bureau that a hurricane was forming and would likely go west into Texas. The Cubans, experts at hurricane prediction, based their forecasts not only on barometric pressure but also on long years observing cloud patterns, wind direction, and the "feel" of the sky, weather wisdom gleaned from four centuries of living in the path of powerful Atlantic hurricanes.

The US Weather Bureau, blinded by a sense of cultural superiority, ignored the warnings coming out of Cuba. American meteorologists instead predicted the "minor" storm would turn north toward Virginia because in their view that's what tropical cyclones were supposed to do. Three days later the storm, by then a Category 4 hurricane with estimated wind speeds of 140 mph (225 kph), leveled the city of Galveston, killing 8,000 to 12,000 people (see figure 17). The townspeople had no idea a major storm was coming. There were no warnings and few evacuations. The Great Galveston Hurri-

Figure 17. Damage from the September 1900 hurricane that struck Galveston, Texas, killing between 8,000 and 12,000 people who had no warning the storm was coming. This is still America's greatest natural disaster. Image from NOAA.

cane, as it came to be known, is still the deadliest natural disaster in American history (Larson 1999).

Today we use satellite images, hurricane hunters, global weather stations, and computer models to forecast hurricane wind speed and direction. But even with modern technology there remains some degree of imprecision, prompting the need to draw "cones of uncertainty" on weather maps. Climate change has further complicated hurricane prediction. Any divinations from the high-tech oracles of forecasting must consider how hotter temperatures are making hurricanes larger, more powerful, and slower moving (therefore more damaging). Tropical cyclones also develop faster, intensifying from tropical storm status to hurricanes more quickly than in the past. This makes a forecaster's job far more nerve-wracking since rapid storm development means less time to issue evacuation orders (Penny 2021). The paths Atlantic hurricanes have taken historically since the mid-1800s are shown in figure 18, including that of the Great Galveston Hurricane, shown as one of the purple lines converging on Galveston Island.

Evil Spirits

The Taino, an Indigenous people native to parts of the Caribbean, gave Europeans the word *hurricane* from their word *hurucane*, meaning "evil spirit of the wind." Hurricanes are officially a category of tropical cyclones. They

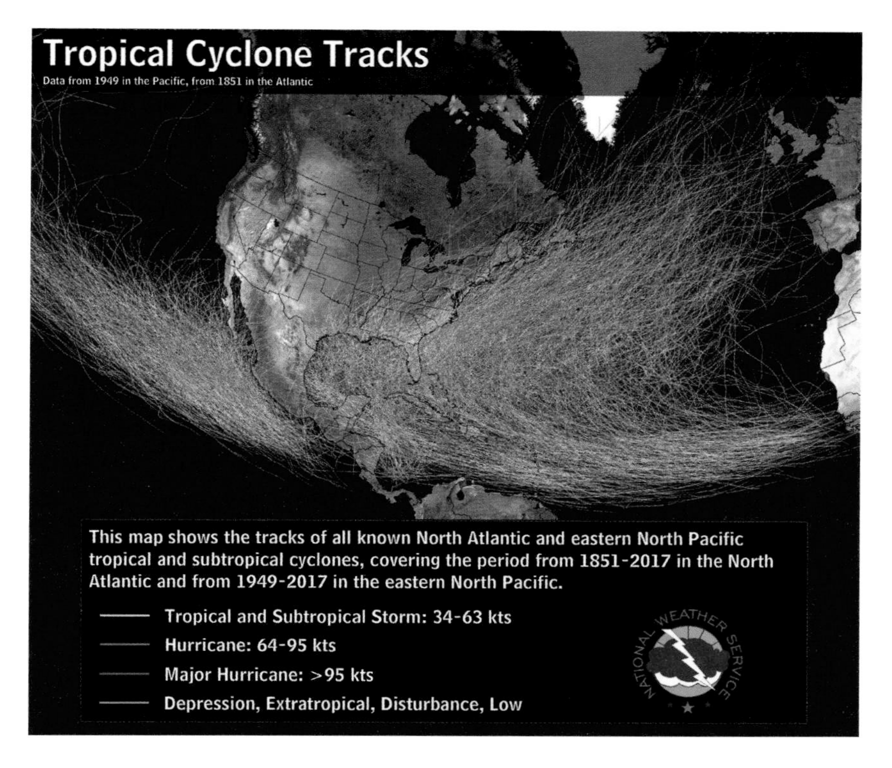

Figure 18. Hurricane tracks in the North Atlantic from 1851 to 2017 and in the eastern North Pacific from 1949 to 2017. Hurricane speed is indicated in knots (kts). Image from the National Weather Service, NOAA.

begin as areas of low pressure over tropical seas where the water temperature is at least 80°F (27°C) and the warm water depth is 150 feet (46 m) or more (Challoner 2014). In the Pacific, hurricanes are known as *typhoons*, in the Indian Ocean as *cyclones* . . . different names for the same evil spirits.

A tropical cyclone forms when hot air rises to be replaced by surrounding air that moves in from all sides toward the center of low pressure. The air moving inward is forced to spin by the Earth's rotation. When the rotating winds exceed 74 mph (119 kmh), a hurricane, complete with a calm, cloudless eye in the center, is born. The winds rotate around the eye counterclockwise in the Northern Hemisphere, clockwise in the Southern Hemisphere (Challoner 2014).

Scientists classify tropical cyclones according to their wind speed and other characteristics as shown in table 2.

Most Americans are familiar with the categories of the Saffir-Simpson Hurricane Wind Scale, even if they don't know the name. The Saffir-Simpson scale categorizes hurricanes according to their wind velocity (see table 3).

Table 2. Tropical Cyclones: Atmospheric Characteristics and Wind Speed

Tropical Cyclone Type	Characteristics	Wind Speed
Tropical Disturbance	A low-pressure area with no closed circulation.	Less than 23 mph
Tropical Depression	Closed circulation with multiple thunderstorms.	Between 23 and 38 mph
Tropical Storm	Circular shape. Damage from heavy rain and storm surge.	Between 39 and 73 mph
Hurricane	Pronounced eye and spiral bands. Damage from heavy winds, storm surge, waves, and tornadoes.	74 mph or more
Major Hurricane	A tropical cyclone with sustained winds of 111 mph or more, corresponding to a Category 3, 4, or 5 hurricane on the Saffir-Simpson scale. Possible catastrophic damage.	111 mph or more

Source: Adapted from NOAA website.

Table 3. The Saffir-Simpson Hurricane Wind Scale

Category	Wind Speed	Damage
1	74 – 95 mph	Very dangerous
2	96 – 110 mph	Extremely dangerous
3	111 – 129 mph	Devastating
4	130 – 156 mph	Catastrophic
5	≥ 157 mph	Catastrophic

Source: NOAA National Hurricane Center. The scale is found here: https://www.nhc.noaa.gov/aboutsshws.php.

It is of course a human construct, a metaphor if you like, unrecognized by Nature. But it does give us a rough measure of the degree of destruction a hurricane can do.

There are factors other than wind speed that influence a hurricane's impact on a coastal community. Beach elevation, bathymetry (the shape and depth of the seafloor), tides, rainfall, the forward speed of a hurricane, coastal vegetation, the location and proximity of the eye, the storm surge, tornadoes, building codes, and, as we learned from the example of the 1900 Galveston storm, adequate time for evacuation are some of the many determinants of a hurricane's destruction. The orientation of the shoreline is also critical in determining the impact of a storm. Places that stick out like Flor-

Table 4. The 10 Deadliest US Hurricanes Since the Beginning of the 20th Century

Ranking	Hurricane	Year	Category	Fatalities
1	Great Galveston Hurricane	1900	4	8,000+
2	Maria	2017	5	2,975
3	Okeechobee Hurricane	1928	4	2,000+
4	Katrina	2005	3	1,200
5	Atlantic-Gulf Hurricane	1919	4	600+
6	New England Hurricane	1938	3	600
7	Audrey	1957	4	500+
8	Florida Keys Hurricane	1935	5	408
9	Miami Hurricane	1926	4	372
10	Grand Isle	1909	4	350

Note: There is some disagreement about precise death tolls, especially for storms from the early years of the 20th century. The fatalities from Hurricane Maria are for Puerto Rico, a US territory. Source: Data from NOAA.

ida, North Carolina, and Nova Scotia are likely to be hit harder than their neighbors. Embayments (shallow bays) like New York Bight are even worse than protruding shoreline segments because they concentrate the storm's energy. The ten deadliest hurricanes in United States history since 1900 are listed in table 4 (NOAA 2021a).

The Pilkey Grandparents' Escape

Orrin Pilkey (Senior) was a brilliant man. He became an engineer and designed bridges, Liberty ships, underwater habitats, and even the boosters for NASA's moon rockets. A man of wit and wisdom, he read the classics and was fond of writing poetry, unusual skill sets for an engineer, to say the least. Yet when it came to dealing with hurricanes, my grandfather was a fool.

In 1969 Orrin and his wife, Betty, lived in a one-story house in Waveland, Mississippi, about half a mile (0.8 km) from the beach. In August of that year, Hurricane Camille serpentined across the Atlantic to make ferocious landfall over Waveland as a Category 5 hurricane. Orrin watched the TV weather reports, carefully calculating where the storm would land. When it became frighteningly clear where it was heading, he and my grandmother picked up their dog and drove away. By then the winds were hissing at hurricane strength, and the roads were flooded. Orrin later recounted seeing in his rearview mirror a tree falling across the road, understanding that their escape would have been blocked had they left a few seconds later. Little did they know at the time that Camille's 20-foot (6 m) storm surge would leave 4 inches (10 cm) of mud on the floor and a 4-foot-high (~1 m) waterline on the walls of their house.

The next day my father, his brother, and I drove down from North Carolina to help with the recovery. The sights of that visit are seared in my memory like a branding iron. Clichéd phrases like *war zone* come nowhere close to describing the aftermath of a major hurricane. All the houses on the beach were gone, leaving only foundations. The same with the next row of houses. And the next. I remember seeing a large tugboat perched upright on a pile of broken lumber that used to be someone's house. Hundred-year-old pine trees stood snapped in half like chopsticks. Mounds of sand and debris were everywhere. And under the debris, though we did not know it yet, lay the bodies of the dead.

My grandparents had been lucky. They made it to Orrin's office at the NASA rocket test site, housed in a stout building designed to survive an accidental rocket explosion. Years later, just before the grandparents moved to Virginia, the family celebrated a final Christmas in Waveland. By then, the town had been rebuilt, new houses constructed on the ruins of the old like some modern-day Troy. Of course we talked about Camille and its devastating destruction (see figure 19). None of us could have imagined that in 2005 an even more deadly hurricane called Katrina would level Waveland again with a 30-foot (9 m) storm surge and that the old family house, which had survived Camille, would be erased from the face of the Earth.

—Charles Pilkey

Lessons from the Pilkey Escape

1. Leave immediately when ordered to evacuate from an approaching storm.
2. Make preparations before the storm is near. In hindsight, while they were dillydallying, the Pilkeys could have gathered important papers and possessions and placed them in high cabinets or in the attic in case of flooding.
3. Choose a good place to wait out the storm. The Pilkeys made a good choice: the administration building at NASA's Saturn Rocket Engine test site (where Orrin Sr. worked) was built especially strong.

After striking the Gulf Coast, the remnants of Camille curved northeastward up the spine of Appalachia, dumping rain in excess of 25 inches (64 cm)

Figure 19. A highway sign along US 29 north near Lynchburg, Virginia, memorializing the disaster caused by Hurricane Camille. The storm came ashore in Mississippi in August 1969 hundreds of miles to the south. Upon arrival in southern Virginia, it produced entirely unexpected catastrophic rainfall and flooding. Photo by Famartin (Creative Commons Attribution-ShareAlike 4.0 International license).

in parts of Virginia. The ensuing landslides on the steep forested mountains near Charlottesville, Virginia, left scars that were still visible decades later (see figure 19).

Mary Edna Fraser Family Escape

Mary Edna Fraser, an accomplished batik artist specializing in coastal scenes, lives in Charleston, South Carolina. In mid-September 1999, Charleston was threatened by Hurricane Floyd, a Category 5 storm when it passed over the Bahamas. Because of its strength, the storm was taken quite seriously, and communities along the South Carolina coast and parts of the Florida and North Carolina coasts were ordered to evacuate.

So Mary Edna packed up a few supplies and snacks and bundled her two small daughters and the family's dog and rabbit into the family car and off they went to Interstate 26 leading northwest to Columbia, South Carolina, normally a drive of about 1 hour and 40 minutes. Her physician husband stayed behind in the local hospital on the assumption that he would be needed to work with storm victims. The escapees quickly slowed to a stopping-and-starting crawl on reaching the densely packed interstate's outgoing lanes. The adjacent incoming interstate lanes into Charleston were completely devoid of any traffic. It was announced that the incoming half of the interstate was reserved for emergency vehicles. After a long haul of more than 9 hours, they arrived in Columbia, having driven along a route with no open gas stations, restaurants, or bathrooms. That night, they stayed safely with friends in Columbia.

It turned out that Hurricane Floyd barely skirted South Carolina but did massive flood damage in North Carolina, killing 51 people, most in riverside towns on the coastal plain, far removed from the shoreline. Had the storm moved into South Carolina as expected, it would have created havoc on the crowded highway.

— Orrin Pilkey

Lessons from the Fraser Family Escape

1. It was commendable that officials ordered evacuation even though it turned out to be unnecessary. *Better be safe than sorry* with such a powerful storm.
2. The family correctly responded immediately and included the family pets.
3. All lanes of interstate highways should be used for escaping traffic.

The Charleston evacuation would have been a disaster if the hurricane had followed its expected path inland. Thousands of motorists could have been trapped. Federal and state officials declared that they had learned a lesson after Floyd and that from then on all lanes of interstate highways would be used for escaping traffic. Unfortunately, this didn't happen in the escape from New Orleans before Hurricane Ida (2021) made landfall. Nor has it happened since.

The Granddaddy of All Evacuations

Perhaps 3 million people evacuated from Houston and parts of East Texas and Louisiana in late September 2005 to escape Hurricane Rita. It was the largest such evacuation in the nation's history. Up to a hundred people may have been killed in this evacuation. Heat stroke killed dozens, and a bus carrying evacuees from a nursing home caught fire, killing 24. Rita was a Category 5 storm as it approached the Texas-Louisiana coast but quickly softened to a Category 3 on hitting land. Adding immensely to the problem was that escapees on Interstate 45 were confined to outgoing lanes only, and the incoming lanes remained empty, a tragic mistake.

Figure 20. The Texas National Guard in flood rescue operations following Hurricane Harvey in 2017. Perhaps this scene would not have occurred if everyone had obeyed the order to evacuate. Photo by Lt. Zachary West, US Army, Texas National Guard.

What Hath Climate Change Wrought?

As mentioned above, tropical cyclones are on average getting bigger and more powerful. Warmer surface waters allow hurricanes to intensify more rapidly. Hurricanes also produce more rain, in part because a warmer atmosphere holds more water vapor (7% more water for every degree Celsius of warming) and also because hurricanes are on average advancing more slowly, allowing more time for rain to fall. This was the problem with Hurricane Harvey in 2017. Not only did Harvey undergo rapid intensification, developing from a tropical depression to a Category 4 hurricane in a mere 48 hours, but it slowed and then stalled as it made landfall in the Houston area, dumping rain in biblical proportions, in some places 24 inches (~60 cm) of rain in as many hours (see figure 20). Rainfall totals in Nederland, Texas, exceeded 60 inches (154 cm), the most for any hurricane in American history. (Noah would have been impressed.) Tens of thousands had to be rescued from the floodwaters. And with a price tag of $125 billion, Harvey became the second-costliest hurricane in US history after Katrina (National Weather Service, n.d.).

September is the cruelest month . . . as far as hurricanes are concerned, for that is the peak of the Atlantic hurricane season (see figure 21). Typhoons can occur all year round, so there is no official typhoon season. The same is true for cyclones in the Indian Ocean. Typhoons are most active from

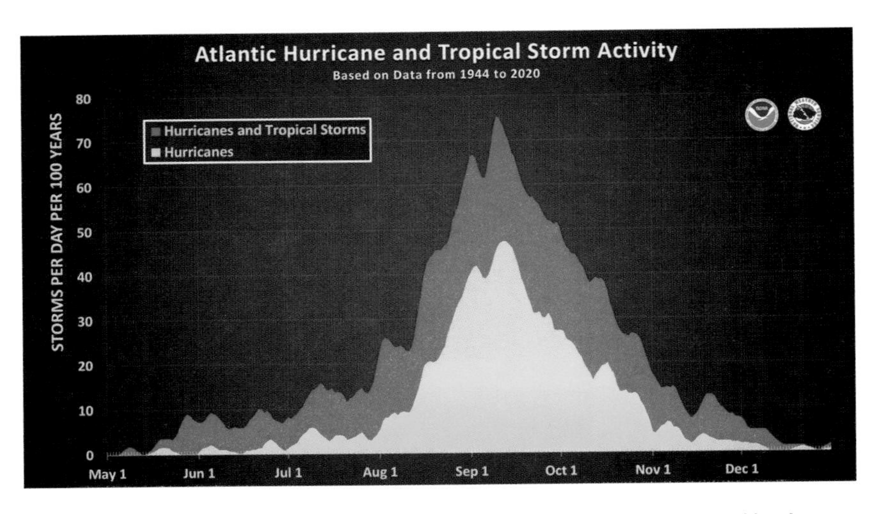

Figure 21. A plot of the monthly occurrence of Atlantic hurricanes as measured by the number of storms over 100 years. September is the peak of the Atlantic hurricane season. Image from National Hurricane Center, NOAA.

summer to fall. Cyclones in the Southern Hemisphere tend to be active in the winter (see table 5).

The effects of climate change sometimes meld into surprising synergies. By midcentury, the sea is expected to rise in most places by a foot (30 cm) or more and perhaps 3 to 8 feet (1.0 to 2.4 m) by century's end (NOAA 2021a). This means storm surges will be proportionately higher even as warmer surface waters fuel stronger hurricanes that in turn spawn higher storm surges. Hurricanes and typhoons used to reach maximum intensity in the tropics. Now, because of warming oceans, the zone where tropical storms reach maximum intensity is shifting out of the tropics toward the poles at a rate of 33 to 39 miles (53 to 63 km) per decade. Put simply, hurricanes and typhoons are moving poleward. Folks living in Canada, Europe, and Down Under should pay heed (Phillips 2014).

The landfall of a tropical cyclone can be devastating, not only for people living on a coast but for marine ecosystems as well. Because coral reefs are generally in shallow water, a powerful storm can smash a reef system, killing individual corals and stressing the fish and other organisms living there. Sometimes oysters, clams, and other salt-loving shellfish are damaged not by winds and waves but by salinity changes caused by massive inundations of freshwater from the heavy rains and flooded rivers that accompany a hurricane's landfall. As the atmosphere heats up and tropical cyclones grow more powerful, damage to wetlands, marine ecosystems, and coastal forests will increase proportionally.

Table 5. Tropical Hurricane and Cyclone Seasons by Region

Region	Season
Atlantic hurricane season	June – November
Eastern Pacific hurricane season	May – November
Northwest Pacific typhoon season	All year
North Indian cyclone season	April – December
Southwest Indian cyclone season	October – May
Australian/Southeast Indian cyclone season	October – May
Australian/Southwest Pacific cyclone season	November – April

Source: WeatherSTEMScholar website, based on NOAA's map of Tropical Cyclone Basins, https://learn.weatherstem.com/courses/wxstem_meteorology_01/module-03/05/06.html.

Hurricanes have a long and deadly history in the Americas, generating floods, wind damage, coastal erosion, power outages, death, and the destruction of coastal ecosystems. Birds and flying insects are especially vulnerable. Strong winds from Idalia, which tore through Florida in 2023 as a Category 3 storm, blew flamingos as far north as Milwaukee (no one saw where the bugs landed).

When Will We Ever Learn?

- In Hurricane Floyd in 1999, evacuation was possible only on the outgoing interstate lanes. Incoming lanes were empty.
- In Hurricane Rita in 2005, evacuation was possible only on the outgoing interstate lanes. Incoming lanes were empty.
- In Hurricane Ida in 2021, evacuation was possible only on the outgoing interstate lanes. Incoming lanes were empty.

What to Do

- **Relocation.** The simplest way to avoid the destructive power of tomorrow's super-cyclones is to get out of harm's way. This could mean a temporary evacuation or a permanent move far from the coast or at a higher elevation. (We recommend the last two.) It also means: don't buy or build a house on a beach or in a flood zone, and do remember that climate change is expanding flood zones and pushing shorelines inland. An up-and-coming choice for safe housing is the dome house. Dome houses are a new concept developed in the Philippines for a low-cost public housing solution in a country

Figure 22. Dome houses in Indonesia. Assuming they are well built, these low-cost EcoShells resist hurricane (and tornado) winds. They have no overhanging ledges or flat surfaces that might catch the wind, but they are not intended to resist flooding. Photo © Monolithic Dome Institute.

with frequent typhoon winds (see figure 22). According to the builder, Texas-based Monolithic Constructors, Inc., the structures can withstand winds up to 400 miles (over 600 km) per hour. The strength of the buildings comes from the fact that there are no ledges like an overhanging roof that the wind will catch and rip away.

- **Pre-storm evacuation (homeowners).** Practice dry-run evacuations with your family and pets. Leave an axe in the attic (just in case). Put together a bag of essential items (see "Bug-Out Bags"). Make sure you have extra water, people food, and pet food in case you get stranded in traffic. Keep your gas tank full during hurricane season. Shop for groceries and other necessities before the crowds rush to the stores.
- **Pre-storm evacuation (city and state governments).** Officials who declare mandatory evacuation must do so more quickly in response to the expected rapid intensification of future storms, including making incoming traffic lanes available for escaping traffic (that is, all lanes become outgoing).
- **Evacuation.** Shut off the gas and power, charge your phone, and unplug everything. Don't evacuate at the last minute (see the escape

stories) when the highways will be congested with evacuees. Don't wait for evacuation orders. Better play it safe, especially if you live in a trailer. The motto for the age of climate change is: Don't wait. Evacuate!

- **If evacuation is not possible.** Before the storm arrives, locate any nurses, doctors, or police personnel in your neighborhood. Before hurricane season starts, learn to be a first responder. Learn first aid, CPR, and so forth. Buy a flare gun to attract the attention of a helicopter. Remember, power and cell service may be out after a major storm. It may be hours or days before the cavalry arrives.
- **Gimme shelter.** Most coastal cities have hurricane shelters. Some of these are Red Cross shelters, but more often than not they are improvised emergency shelters in schools and churches. A godsend for the poor perhaps, but not so much for pets. Most shelters can accommodate a limited number of evacuees but don't accept pets. We do not recommend this, but IF the motels are full, you can temporarily shelter in your car (pets and all) at a rest area or parking lot, provided you have driven a safe distance from any storm surge, heavy winds, or tornadoes and are not in the future path of the storm. There are companies that sell tornado and hurricane shelters in compliance with FEMA (Federal Emergency Management Agency) standards. Make sure the shelter you buy or build is elevated above the height of the maximum local storm surge. On the Gulf Coast this means higher than 30 feet (9 m). Visit FEMA's website (fema.gov) to learn about safe rooms and storm shelters.

Tornadoes

Toto, I've a feeling we're not in Kansas anymore.
—Dorothy in *The Wizard of Oz*

The Nature of the Beast

To the scientist perusing computer models in the comfort and safety of an academic office, a tornado is an enigmatic and fascinating vortex of rotating air. To the survivor crawling through the rubble of a ruined home, a tornado is a beast, a living, breathing monster leaving an indelible path of sorrow across the landscape of memory.

Sometimes a tornado's aftermath can seem whimsical, as when a 2011 storm deposited mail from Cullman, Alabama, on people's yards in Cleveland, Tennessee, a distance of more than 150 miles (241 km). Or miraculous, as when a cat emerged unharmed after being buried for nine days in the ruins of a Mayfield, Kentucky, office building, leveled by an EF4 tornado (table 7 displays the Enhanced Fujita Scale categorizing tornado strength). But more often than not, the passage of a tornado is marked by suffering and terrible loss.

The United States has around 1,200 tornadoes per year (four times as many as the rest of the world combined) that on average kill 80 people and

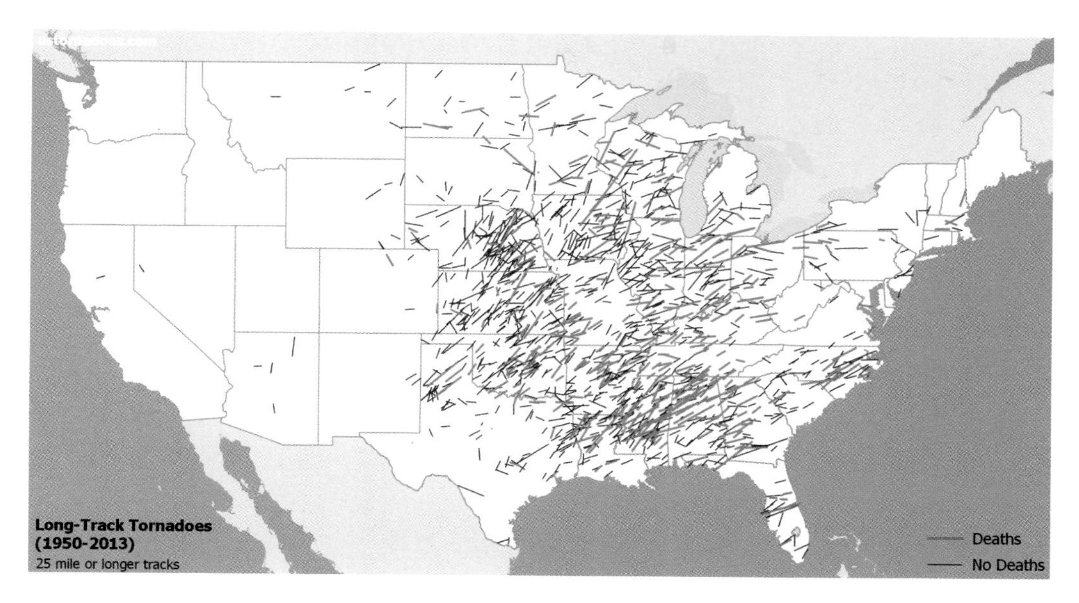

Figure 23. Map showing long-track tornadoes (more than 25 miles [40 km] long) in the United States between 1950 and 2013. The red lines indicate tornadoes in which deaths occurred. Blue lines indicate tornadoes that resulted in no deaths. Image from NOAA.

injure 1,500 (see figure 23). Some can be quite powerful. The 1999 Bridge Creek–Moore tornado in Oklahoma City had winds measured at over 300 mph (483 kmh), the highest recorded wind speed on planet Earth.

Tornadoes come in a variety of shapes and sizes. Those over water are known as *waterspouts*. They are generally weaker than the ones on land and are not included in the tally of American tornadoes. Multiple-vortex tornadoes have two or more funnels revolving in close proximity around the same vortex (see figure 24), not to be confused with tornado families or outbreaks, which are different tornado systems often parallel to each other and linked to the same storm system. Other tornado-like phenomena include gustnados (caused by rapid thunderstorm downbursts), dust devils (whirlwinds common on hot days in the desert Southwest), and firenadoes (fiery funnels erupting from large, intense wildfires).

Tornadoes today are ranked using the Enhanced Fujita Scale (EF Scale) which is based on the amount of wind damage. It is a modified version of the original, no-longer-used Fujita Scale (based on wind speed) developed by Theodore Fujita in 1971 (see tables 6 and 7).

The EF Scale is based on estimated wind speeds over a 3-second interval. The wind velocity is determined by observing structural damage to any of 28 damage indicators compiled by the National Weather Service. This includes damage to homes, malls, businesses, schools, trees, and so forth.

Figure 24. Twin tornadoes on the Great Plains, September 28, 2010. The most tornadoes spawned by a single storm system occurred in the late April 2011 Super Outbreak, when 362 tornadoes were spotted from Texas to southern Canada, with Alabama being hit the hardest. Such outbreaks are becoming more common. Tornadoes are staying on the ground longer, and Tornado Alley may be moving eastward. Photo from NOAA Legacy Photo; OAR / ERL / Wave Propagation Laboratory.

The EF5 category has no stated upper limit, purposely designed to keep the media from exaggerating wind speeds. See figure 25 to appreciate the kind of damage a powerful EF4 tornado can inflict on a community.

In the December 29, 2021, *New York Times*, Christopher Flavelle argues that the US toll in damage and deaths reflects poor human decisions as much as the force of the tornado wind. Experts note that well-built "safe rooms" offer "near absolute protection," and following well-known principles of high wind construction will keep buildings from flying apart. Efforts to incorporate these into modern construction codes have repeatedly failed because of the building industry's concern with high costs. The lack of protective regulations in tornado country boils down to money. Ironically, the principles of high-wind construction are usually required for buildings located along hurricane-prone shorelines.

Tornadoes and Climate Change

A reasonable assumption might be that global warming will lead to stronger and more frequent tornadoes. But links between climate change and tornadoes are not so clear, in part because the statistical record of past tornadoes

Table 6. The Fujita Scale

Scale	Character	Estimated Winds	Description
F0	Weak	40 – 72 mph	Light Damage. Some damage to chimneys; branches broken off trees; shallow-rooted trees uprooted; sign boards damaged.
F1	Weak	73 – 112 mph	Moderate damage. Roof surfaces peeled off; mobile homes pushed off foundations or overturned; moving autos pushed off road.
F2	Strong	113 – 157 mph	Considerable damage. Roofs torn from frame houses; mobile homes demolished; boxcars pushed over; large trees snapped or uprooted; light objects become projectiles.
F3	Strong	158 – 206 mph	Severe damage. Roofs and some walls torn from well-constructed houses; trains overturned; most trees in forested area uprooted; heavy cars lifted and thrown.
F4	Violent	207 – 260 mph	Devastating damage. Well-constructed houses leveled; structures with weak foundation blown some distance; cars thrown; large missiles generated.
F5	Violent	260 – 318 mph	Incredible damage. Strong frame houses lifted off foundations, carried considerable distances, and disintegrated; auto-size missiles airborne for several hundred feet or more; trees debarked.

Note: Ted Fujita (University of Chicago) and Allen Pearson (National Severe Storms Forecast Center, now known as the Storm Prediction Center) developed the Fujita Scale for measuring the strength of tornadoes in 1971 and updated it two years later.
Source: NOAA and National Weather Service.

Table 7. Enhanced Fujita Scale

Rating	3-Second Gust (mph)	Damage Description
EF0	65 – 85	Minor
EF1	86 – 110	Moderate
EF2	111 – 135	Strong
EF3	136 – 165	Severe
EF4	166 – 200	Devastating
EF5	Over 200	Incredible

Note: The Enhanced Fujita Scale (EFS) replaced the Fujita Scale in the United States in 2007. The EFS provides a more accurate match of tornado wind speeds to damage on the ground. Source: NOAA and National Weather Service.

Figure 25. The ruins of the now-famous Mayfield Consumer Products candle factory in Mayfield, Kentucky, where 100 people were working when the December 10–11, 2021, tornadoes struck the building. The building was completely flattened, and eight people died. The lesson here is that buildings in Tornado Alley must be built using high-wind construction standards as is already done for buildings in Hurricane Alley. In addition, storm shelters or "safe rooms" should be constructed of sufficient size to shelter the likely number of people who will use them. Photo © Ryan C. Hermens, *Lexington Herald-Leader*.

is incomplete and also because current climate models don't work well with small-scale, ephemeral weather events like tornadoes. However, some apparent patterns of tornado activity are starting to emerge.

While the annual number of tornadoes has not increased over the past century, the number of days with outbreaks (multiple tornadoes birthed from the same storm) has. There is growing evidence that tornadoes are on average wider and stay on the ground longer than in the past. Moreover, the area of frequent tornado formation in Kansas and Oklahoma, known as Tornado Alley, is apparently moving east toward the Mississippi and Ohio River valleys. In 2021 Minnesota had its first recorded tornado in December (a total of 16 were reported), suggesting that a warming climate may extend the normal tornado season into winter.

REMEMBER:
A tornado watch means "Be prepared."
A tornado warning means "Seek shelter."

What to Do

- **Communities should upgrade** building codes to accommodate changing tornado patterns.
- **Choose a storm room** (preferably with no windows), usually a bathroom, basement, or a space under the stairs, to ride out the storm. Take a solar radio and/or crank weather alert radio into the storm room for weather updates if the power goes out. Make sure you have enough food, water, and medicine for long-term power outages. Do tornado drills with your family.
- **Consider purchasing a storm shelter.** These are available commercially. The cost of a substantial family shelter starts at around $2,000. Or make a safe room following the guidelines of the Federal Emergency Management Agency (FEMA); see the page "Safe Room Publications and Resources" on FEMA's website (https://www.fema.gov/emergency-managers/risk-management/safe-rooms/resources).
- **If driving toward a tornado**, don't stay in the car (unless absolutely necessary). It is better to exit the car and lie flat in a ditch or low area as far away as possible from your car. Don't take shelter under an overpass. If you can't evacuate from the car in time, keep the engine running to enable deployment of airbags. Stay low, keep your seatbelt on, and protect your head.
- **If in a house during a tornado**, stay in the most central and lowest part of the house, away from windows. Don't open any windows. Try to keep as many walls, doors, and other barriers between you and the outside as possible. Wear a helmet or wrap your head and body with mattresses, pillows, blankets, and the like.

Heat

The heat is on, on the street
Inside your head, on every beat
And the beat's so loud, deep inside
The pressure's high, just to stay alive
'Cause the heat is on
—Harold Faltermeyer and Keith Forsey,
"The Heat Is On"

A Silent Killer

Heat waves kill more people in the United States than any other extreme weather hazard. When it gets too hot, the human body begins to fall apart, the cardiovascular system collapses, and the kidneys start to fail. Heat is a sly killer that silently creeps up and takes life. The very old and the very young are most susceptible, but we should all be prepared for increases in the occurrence and intensity of heat events.

Wet-Bulb Temperatures

A heat wave is usually defined relative to the typical weather in an area and the normal temperatures for the season. So the threshold for a heat wave in Tucson, Arizona, is higher than the threshold for Seattle, Washington. The heat index, also known as the *apparent temperature*, is what the temperature feels like to the human body when relative humidity is combined with the air temperature. This defines for us how comfortable we are with the temperature

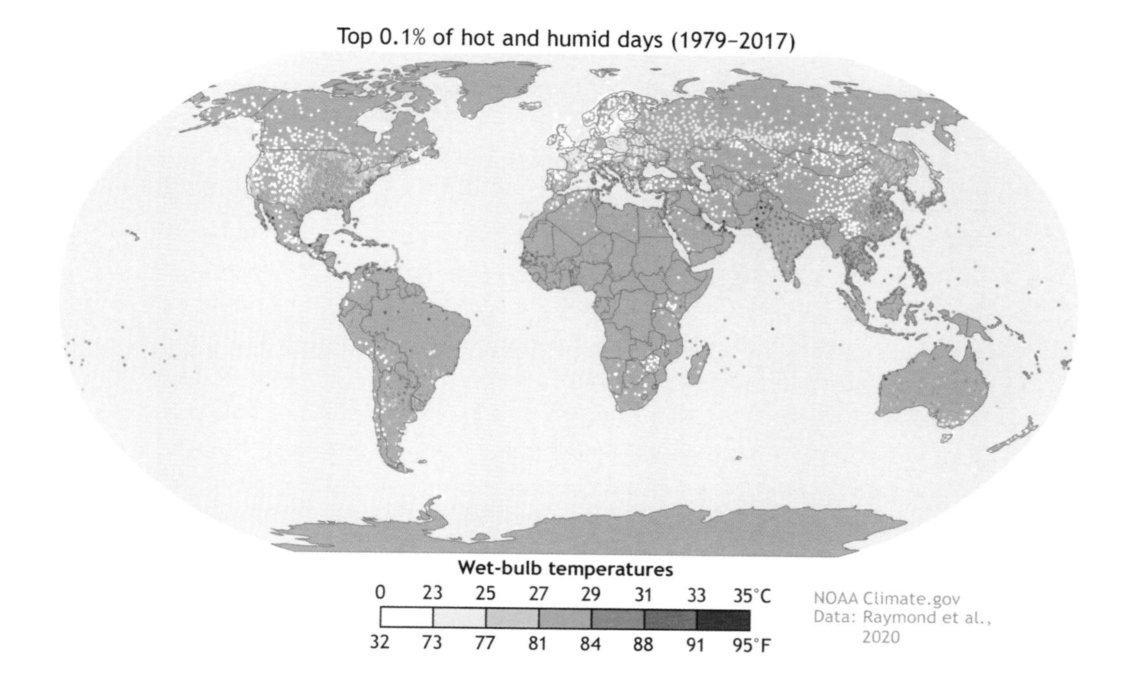

Figure 26. This map shows areas of extreme wet-bulb temperatures (1979–2017). Darker colors show the more severe occurrences. Scientists believe that humid heat is a vastly underestimated climate-change-related health risk. Map from NOAA, climate.gov, based on data from Raymond et al. (2020).

of the day. But the increases in extreme heat have introduced a new term to our climate-change vocabulary—*wet-bulb temperature*—the most important concept related to climate change that no one has ever heard of.

A wet-bulb temperature is a measure of the air temperature at 100% humidity. Scientists used to calculate wet-bulb temperatures by wrapping a thermometer in a wet towel and swinging it around. Now they use electronic instruments at weather stations. The limit for human survival has been identified as exposure to a sustained wet-bulb temperature of 95°F (35°C) (about the temperature of human skin). A wet-bulb temperature above 95°F (35°C) means certain death after only a few hours, regardless of how healthy a person is, no matter whether that person is sitting in the shade nor how much water they are drinking (Buis 2022).

Scientists once theorized that toward the end of the century, global warming might cause heat waves with wet-bulb temperatures of 95°F (35°C). But such temperatures are already happening, decades sooner than expected. Since 2005 wet-bulb temperatures of 95°F (35°C) have occurred nine times (albeit only briefly) in Pakistan and parts of the Persian Gulf (see figure 26).

According to NASA, within 30 to 50 years wet-bulb temperatures will reach or exceed the critical threshold in South Asia, eastern China, the Red Sea, the Persian Gulf, Brazil, and American states like Arkansas, Iowa, and Missouri (Buis 2022). The heat is on.

A new Penn State study put the threshold of human survivability much lower than previous estimates. According to this study, 87.8°F (31°C) is the upper wet-bulb temperature limit for the human body. This is the equivalent of 105°F (40.6°C) at 50% relative humidity or 100°F (37.8°C) at 60% relative humidity. Dr. Larry Kenney, who led the study, concluded "Our results suggest that in humid parts of the world, we should start to get concerned—even about young, healthy people—when it's above 31°C [87.8°F] degrees wet-bulb temperature" (quoted in Tanenbaum 2022). If the Penn State data hold true, then heat waves are a greater existential threat than was previously thought.

It should be noted there are temperature/humidity limits for wildlife and livestock as well. The 2021 heat wave in British Columbia killed hundreds of thousands of farm animals. Placental mammals have a similar tolerance range for heat and humidity as humans, but with one significant difference. Deer, bison, monkeys, and squirrels can't buy air conditioners (Huber 2022).

When exposure to heat is high enough to raise the body's core temperature, heat stress illness (cramps, exhaustion, and heatstroke) can occur. With heat cramps, an individual experiences muscle pain or spasms. Heat exhaustion can cause dizziness, a weak pulse, nausea, and fainting. The most severe heat-related illness, heatstroke, occurs when a person's temperature increases above 103°F (39.4°C). Increased daily air temperatures or periods of extended high temperatures have also been shown to increase cardiovascular mortality, respiratory mortality, and heart attacks.

Disproportionate Impacts

The 1995 heat wave in Chicago lasted five days, killing perhaps 739 people. By the afternoon of day 1, July 13, temperatures hit 104°F (40°C) at O'Hare International Airport. With the heat index, it felt like 125°F (51.7°C). This event first brought wide recognition to the inequities of death during heat events. Eric Klinenberg, author of *Heat Wave: A Social Autopsy of Disaster in Chicago*, documented the heat-related deaths in Chicago, noting the social conditions that contributed to the deaths. Klinenberg (2015) found that the deaths closely mirrored the "ecology of inequality" in the city, concentrated as they were in low-income neighborhoods, among the elderly, and in the city's most violence-prone areas.

Death rates during heat waves are generally highest for older people and those who are homeless, socially isolated, and without air-conditioning. This

Table 8. A Short History of Heat Waves

Europe, heat wave of 1757	Records describe this as the hottest summer in Europe for 500 years, until the summer of 2003.
Argentina, "Week of Fire," 1900	Occurred in February and lasted a week, with temperatures of up to 99°F (37°C). More than 478 deaths.
The North American heat wave of 1936	This heat wave occurred during the Depression in the Dust Bowl (southern plains region from Texas to Nebraska) after one of the coldest winters ever recorded.
The Chicago heat wave of 1995	Estimated 739 deaths. Mostly over 5 summer days. Today would be called a *heat dome*. Most of the deaths were poor, elderly residents who could not afford air-conditioners and did not open windows or sleep outside for fear of crime.
The European heat wave of 2003	Caused an estimated 70,000 deaths; 14,000 deaths in France alone. Most who died were very young children, the chronically ill, and the elderly living alone.
The Russian heat wave of	55,000 deaths. Deaths in Moscow averaged 700 a day 2010 (twice the average number). The event crippled the Russian wheat harvest, leading to economic losses in the billions.
Australia, "Angry Summer" of 2012 – 13	Numerous weather records were broken over a 90-day period: the hottest day ever recorded, the hottest January, the hottest summer average, and a record 17 days in a row when the whole continent averaged above 102°F (39°C).
Siberia, summer of 2020	In June 2020 a sweltering 100.4°F (38°C) was recorded in the Russian town of Verkhoyansk. This is the highest temperature ever recorded above the Arctic Circle.
Japan, 2022	The worst heat wave in 150 years, with the hottest temperatures recorded since 1875.
Northern Hemisphere, summer of 2023	July 2023 was the hottest month ever recorded on Earth. Cities in the United States, Europe, and Asia logged record-breaking high temperatures.

was the case in France during the 2003 European heat wave. Of the estimated 14,000 heat-related deaths in France (some put the number closer to 20,000), many were elderly people living alone. Few had air conditioners (see table 8). In the past 20 years, heat-related mortality among people over 65 years of age has increased by more than 50% (Atwoli et al. 2021, 1134). Heat also disproportionately impacts children, pregnant women, and people with chronic health conditions. For obvious reasons, outdoor workers (e.g., in construction and agriculture) are also vulnerable to extreme heat events.

Impact of Heat on Infrastructure

Heat takes a heavy toll on the power grid. Hotter weather can lower the amount of energy that power lines transmit. Increased demand for electricity to run air conditioners (which are a vital public health tool to those with access) can cause the power grid to falter or fail. Additionally, high temperatures can limit power production when the water used to cool power plants becomes too warm. Heat can lead to droughts, which in turn cause water levels to fall too low for hydropower plants to operate fully (see "Drought" in "Water"). Rolling blackouts are becoming a norm as a reaction to demands on power grid infrastructure that is not designed to take the strain.

Heat also leads to disruption of transportation, travel, shipping, and manufacturing. Heat expands concrete, and during extreme heat waves, roads buckle and break. Steel drawbridges expand, hindering their ability to open and close. Train tracks form "sun kinks" as the metal bends, increasing the risk of derailment (e.g., the 2019 European heat wave where widespread train cancellations across the continent were necessary). Planes can struggle to fly in high-heat conditions, and extreme temperatures can reduce the allowable weight of aircraft for takeoff.

As we look into the future, engineers might have to consider new standards for infrastructure (see "Green Cities" in "Space"), and people will need to become more self-reliant to deal with blackouts and other interruptions to their lives.

Heat Domes

A major atmospheric feature leading to some of the worst heat waves globally is the heat dome, a deadly event that is bound to occur with increasing frequency. A heat dome happens when a high-pressure system covers a large area for prolonged periods of time, trapping very warm air under a stable atmospheric dome (see figure 27). Under the dome, hot air rises until it is blocked by the dome's lid. This causes the air to sink to the earth, warming as it does. In a dome event, there is no relief brought by cooler overnight temperatures. The high-pressure system stalls and blocks other fronts from moving through and pushing it on. A dome event can thus become a kind of self-perpetuating atmospheric cap.

Heat domes are distinguished from heat waves by their strength, wide geographic scope, and persistence. The physiological stress of prolonged and unbroken extreme heat is well documented. The heart works harder to move blood around so that a person can dispel heat through the skin. The face becomes redder as the blood vessels open, trying to release heat. Overnight cooling is critical to give the vascular system a break. But heat domes do not bring the relief to let a body recover.

These events trigger public health emergencies, resulting in excess mortality. A heat dome over northwestern India starting in April 2022 threat-

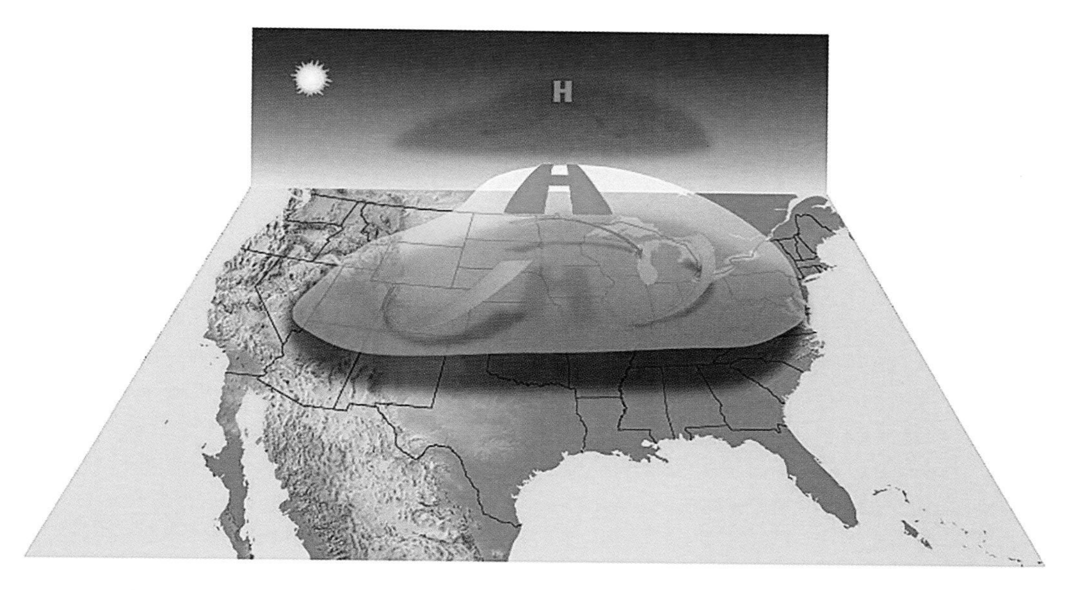

Figure 27. A heat dome is formed when atmospheric pressure stalls in place and traps heat at the Earth's surface, acting like a lid on a pot of boiling water. Image from NOAA (2021a).

ened a billion people with bad air and reduced crop yields. Scientists predict heat domes in India and Pakistan will increase in frequency and severity, rendering much of the Indian subcontinent uninhabitable for humans.

The Pacific Northwest Heat Dome

In 2021 a Pacific Northwest heat dome formed over British Columbia, Washington, and Oregon, sealing in unbearable heat. This occurred at a time when COVID-19 was keeping people indoors and isolated, largely working and schooling from home. More than three consecutive days of highs in the triple digits occurred, with record temperatures in both Portland (116°F/47°C) and Seattle (108°F/42°C). Perhaps as many as 1,400 people in Canada, Washington, and Oregon died.

The heat sparked numerous wildfires in the US Northwest and in British Columbia. The small western Canadian town of Lytton had seen three days of record-breaking heat. On June 30, temperatures in Lytton rose to 121°F (49.4°C), a record high for Canada. The next day a wildfire overran Lytton and burned 90% of the town to the ground. The heat also damaged the road and rail infrastructure, forced closures of businesses, affected marine ecosystems, and disrupted cultural events.

The Washington State Heat Dome

I live near Olympia, Washington, and survived the three consecutive days and sleepless nights of triple digits brought by the infamous Pacific Northwest heat dome of 2021. I grew up in the South and since moving west have long relished the lack of high temperatures and humidity in the Pacific Northwest. I do not have air-conditioning in my house. Air-conditioning exists sparingly in the Pacific Northwest — it has simply not been necessary for feeling comfortable in the heat of summer . . . until recent years.

I had plenty of notice, though in retrospect I was naive about the terminology being used. I heard the words *heat dome* for a week on the news and weather reports before it arrived. I had no idea what that meant. By the time I realized what a heat dome was, on day 1, there were no cooling fans or portable air conditioners to be had. By the second day, there were no vacancies in hotels anywhere as people sought out air-conditioning. Neighbors, friends, and families — we all reached out to each other frequently to share experiences and tips and to make sure we were all okay. My poor cats were miserable and very lethargic. They barely moved. We didn't use the stove or the oven. We survived by taking frequent short, cold showers.

The most vivid part of the experience for me was that nighttime temperatures brought no reprieve (my normal coping mechanism for heat). I carefully covered every window in my house, using double layers of blocking fabric on top of curtains. For relief, I started off using numerous fans strategically placed to push air (where to stage and when to move the fans became an obsession). By the second day, I had to use most of the fans to blow air across the basement floors, as the floors were sweating (condensation was forming) and water was pooling.

By the last day, I recognized that the effects of heat had slowly, imperceptibly crept up on my body and mind. I could feel that my internal body temperature was elevated. I felt ill and dullminded. Today I am better prepared for the heat waves and domes, but I utterly hope I never have to live through another dome event.

— Linda Pilkey-Jarvis

The Pacific Northwest heat dome was considered a one-in-a-thousand-year event and was the most intense one ever recorded in that part of the world. In the summer of 2023, an equally intense heat dome formed over Texas, Mexico, and parts of Oklahoma and Louisiana. San Angelo, Texas, recorded its highest temperature ever at 114°F (45.6°C). Highways buckled under the heat, and the state of Texas's power company, scrambling to avoid a grid collapse, urged people to voluntarily cut back on power usage. In July,

Phoenix suffered a record 31 consecutive days of high temperatures above 110°F (43.3°C). The heat dome extended as far south as Mexico City, which recorded its hottest June day ever at 92.5°F (33.6°C). Those numbers may not impress, until you consider that the city stands 7,200 ft (2,195 m) above sea level and is usually pleasantly cool in June.

The heat dome over the American Southwest was but one of a multitude of extreme heat events occurring simultaneously around the world, making July 2023 the hottest month ever recorded on Earth. Events like the heat domes in 2023 and 2021 should be a strong warning. In a time of rapid climate change, we can expect more heat waves and heat domes and higher wet-bulb temperatures. As individuals we can protect ourselves by preparing now, but we need concerted effort from local, state, and national governments to educate people on the dangers of heat, alert us when temperatures creep too high, and offer ways to survive a world that's likely to grow hotter for a very long time to come.

What to Do

- **Two important tips** are to stay informed of the forecasts and keep your body hydrated.
- **Go to a cooling center** when they are open. Cooling centers are a health intervention used during extreme heat to cool us down. Libraries, town halls, schools, or shopping malls are open to the public, as well as outdoor locations in water spray parks, pools, and parks.
- **Share facilities, supplies, and plans** with neighbors. Those with low incomes and underlying health conditions are the most vulnerable. This is especially important in a heat dome.
- **Create an oasis.** Urban heat islands raise temperatures and retain much of the heat at night. Plant trees for shade.
- **Air conditioners are the main defense**, but often power will be out due to excess numbers of operating air conditioners or from storm damage. Not all can afford air conditioners. Also, air conditioners can increase carbon emissions if the electric grid is powered by fossil fuels.
- **Install ceiling fans and heat pumps.** Both perform well to keep you cool (though less so at very high temperatures).
- **Beware of heat from appliances.**
- **Be cool.** Take a cool shower, cool sheets in the freezer, dampen sheets, use ice packs, stay in the shade, open windows on breezy days, designate a cool room.

- **Exit hot cars.** Don't leave kids or pets in parked cars, even if the windows are open.
- **Drink plenty of water** and eat smaller and more frequent meals. Take breaks, and wear light-colored, loose-fitting clothes. Make sure kids and pets stay cool and drink enough water.
- **Sleep low** (heat rises).
- **Work outdoors in the mornings and evenings only.**

Bad Air

Men come and go, cities rise and fall, whole civilizations appear and disappear—the Earth remains, slightly modified.—Edward Abbey, *Desert Solitaire: A Season in the Wilderness*

Air Pollution Kills

Every day, outdoor air pollution (mostly from fossil fuels) kills 11,000-plus people around the world (see figure 28). The number is higher if you include deaths from indoor pollution. Many of the dead are elderly with preexisting health problems—cardiovascular issues, cancer, asthma, and the like. Most live in China and India. In the United States, polluted air kills over 200,000 every year, averaging more than 500 deaths each day. Air pollution is more than just an environmental irritant. It's a killer (Roser 2021).

Scientists have long warned us about the hazards of polluted air, which is why Congress passed the Clean Air Act of 1963, and the Nixon administration established the US Environmental Protection Agency (EPA) in 1970. By EPA estimates, the Clean Air Act and its amendments have prevented millions of premature deaths for Americans (230,000 in 2020 alone). Cleaner air and less polluted water mean healthier lives, fewer school-day losses, fewer emergency room visits, and fewer premature deaths (EPA 2021b).

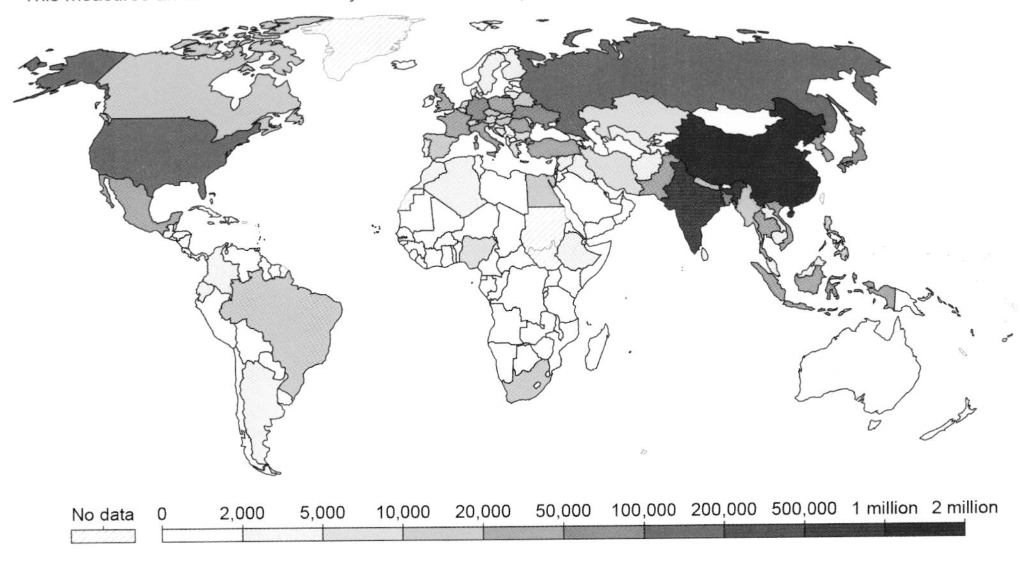

Figure 28. Air pollution deaths from the burning of fossil fuels in 2015. Note that mortality is highest in India and China. Image from Lelieveld et al. (2019). OurWorldInData.org/air-pollution (CC BY).

By 2020 the Trump administration had emasculated the EPA, and while President Joe Biden was able to roll back some of the environmental deregulation imposed by his predecessor, the 2022 decision by a conservative Supreme Court to curtail the EPA's ability to limit carbon dioxide (CO_2) emissions will make it much harder for government to mitigate climate change and will lead to more deaths across the country, especially in poor communities. Climate change does more than just heat the air. It makes the air dirtier, more polluted, and deadlier from ground-level ozone and increased levels of dust and other particulate matter. It also likely makes us stupider.

Aerosols

A polluted atmosphere is one that contains gases, smoke, toxic chemicals, dust, and aerosols in harmful amounts. Atmospheric warming associated with climate change will increase all of the above. Aerosols are fine liquid or solid particles suspended in air. Water vapor is an example of an aerosol. So are pesticides, dust, sulfates, nitrates, sea salt, ozone, and smoke from coal plants, automobile exhausts, and wildfire smoke.

Greenhouse gases (GHGS) (water vapor, CO_2, methane, ozone, etc.) warm the atmosphere by trapping and emitting heat from the sun. But most aerosols have the opposite effect. They cool the atmosphere by reflecting or scattering sunlight into space, thus countering the warming effect of CO_2 and its greenhouse gas brethren. Aerosols have actually slowed the pace of global warming since the inception of the Industrial Revolution, though scientists disagree about how much hotter our planet would be if aerosols hadn't cooled things off. In any case, aerosols are short lived, lasting days or weeks, while CO_2 can remain airborne for centuries and methane a decade or more. Methane absorbs more solar energy than CO_2 and so heats the atmosphere more, but it has a shorter atmospheric lifespan; after 10 years or so methane breaks down into carbon dioxide and water vapor. The warming effects of GHG emissions thus far outweigh the cooling properties of aerosols (Wallace-Wells 2020).

The ability of aerosols to reflect sunlight has prompted some to propose dropping sulfates or sea salt in the upper atmosphere to cool the planet, a kind of geoengineering known as *solar radiation management*. The engineering will certainly work, but as with any new, untested technology, there will be unforeseen consequences. Reflecting sunlight into space could impact photosynthesis. And since aerosols cause cloud formation by providing nuclei around which water vapor condenses, precipitation patterns may change. But even if the consequences of managing solar radiation are minimal, ocean acidification will continue unabated as more industrial CO_2 is drawn into the sea. Engineering a cooler climate with aerosols should be seen as a temporary expedient, buying time for humanity to curb its carbon emissions.

Climate Change and Air Pollution

Ozone, particulate matter, and other air pollutants are interlinked with climate change through complex physical, chemical, and photochemical interactions, some natural and some human engineered.

Ozone (O_3) is an aerosol that happens to be a key component of smog. It is also a GHG. In the upper atmosphere, naturally occurring ozone protects life on Earth from lethal doses of ultraviolet rays. But ground-level ozone is a health hazard, particularly dangerous for people with cardiovascular problems. Ozone forms when nitrous oxides and other emissions from fossil fuel combustion interact with sunlight. When inhaled, ozone causes chest pains, sore throats, coughing, increased asthma, decreased lung capacity, and in some cases death. It also damages crops and forests by chemically "burning" leaves. Soybeans, peanuts, and cotton are especially susceptible to

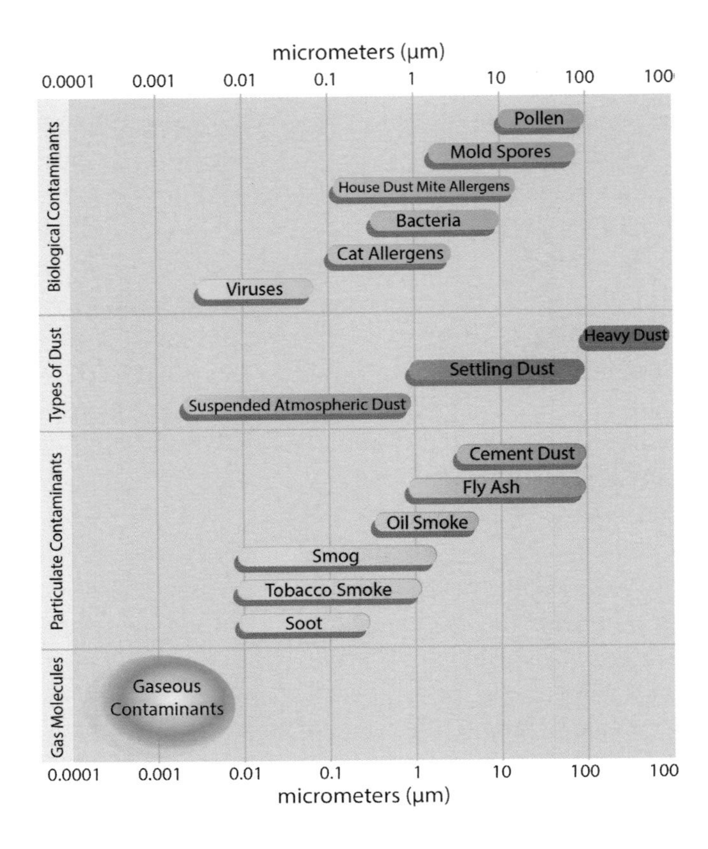

Figure 29. Different kinds of airborne particles as measured in micrometers or microns. The micron is a unit of length equal to one-millionth of a meter. Spider silk is 3–8 microns wide. Image by Mieszko the first (Creative Commons Attribution 3.0 Unported license. https://commons.wikimedia.org/wiki/File:Airborne-particulate-size-chart.jpg).

ozone damage (USDA 2016). Higher temperatures and heat waves increase ozone concentrations, which is why cities commonly have ozone alerts in the summer.

Particulate matter (or particle pollution) is a kind of aerosol made of small particles that are measured in microns. One micron (µm) is one millionth of a meter or about the size of most bacteria. Human hair ranges in width from 20 to 200 microns. Larger particulate matter like soot and smoke is visible to the naked eye (see figure 29). The smaller stuff you can't see is the most dangerous form of air pollution, especially for the $PM_{2.5}$ particles (meaning those that are 2.5 microns or less in size). Particles that small can penetrate deep into the lungs, the brain, and the blood, potentially causing heart attacks, lung cancer, chronic obstructive pulmonary disease (COPD), dementia, and premature death in its various guises.

Examples of particulate matter and their links to climate change are listed below:

- **Dust.** "Black blizzards" were what the farmers of the Great Plains called the dust storms of the 1930s Dust Bowl years made famous by

Figure 30. Dust storm approaching Stratford, Texas, back in the Dust Bowl days of the 1930s. During the severe dust storms of that era, people covered their faces with masks and sealed their windows and doors as best they could. Dust is a serious health hazard for those with asthma and other health problems. Photo from NOAA.

John Steinbeck's *The Grapes of Wrath* (see figure 30). Climate change is causing drought, desertification, wildfires, and deforestation, allowing winds to pick up more dust and sand from the soil. The numbers of dust and sand storms are increasing worldwide. The black blizzards may be making a comeback (Marot 2022).

- **Pathogens.** Viruses, harmful bacteria, and fungal spores as well as toxic metals, pesticides, and herbicides can be transported long distances with other particulate matter by wind. Valley fever (coccidioidomycosis), a fungal infection afflicting humans and livestock in the western United States and Mexico, is spread by wind, often when dust storms pick up fungal spores from the soil. Valley fever was recently found in Washington State, which may indicate the fungus is moving north with a warming climate (CDC 2020b).
- **Allergens.** Warmer temperatures, longer pollen seasons, and higher levels of CO_2 are spurring increased pollen production in plants like ragweed, oak, and cedar (Sommer 2022).
- **Smoke pollution.** Wildfire smoke contains a mixture of gases, liquids, and solid particles. Short-term health issues (for humans, livestock, and wildlife) range from burning eyes and runny noses to aggravation of existing heart or lung diseases. Even worse, particulate

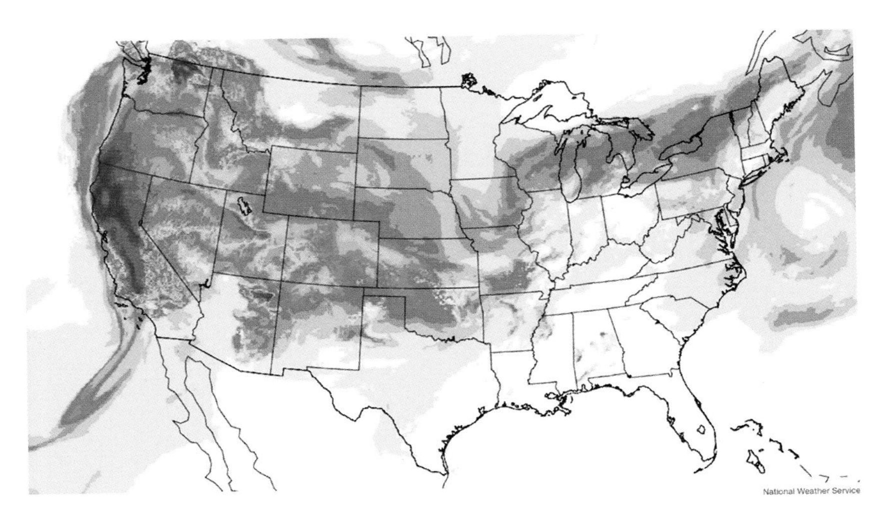

Figure 31. This map shows the distribution of near-surface smoke that traveled across the United States from the 2018 fires in Northern California and British Columbia. The smoke created a hazard for asthma sufferers and provided beautiful sunsets for New Yorkers. Image from NOAA.

matter in smoke can enter the bloodstream and travel to other organs, causing widespread inflammation. Because smoke can travel long distances, even people thousands of miles away from the fires feel its effects. Smoke from recent western wildfires traveled nearly 3,000 miles (4,830 km) to the East Coast, giving New York City hazy skies and beautiful, tomato-red sunsets (see figure 31).

- **Smoke-tainted crops.** Devastating wildfires have burned major wine-producing regions in the United States, Greece, Australia, and Spain. Even fires distant from a vineyard pose a threat to the industry because wildfire smoke alters the smell and taste of wine, effectively ruining some of the world's prized vintages. The term *smoke-tainted* has been coined to describe this result. Grapes, particularly premium grapes for high-end wines, grow best in a temperate climate and therefore are sensitive to climate change, including smoke damage. Smoke and ash are also prolonging the drying time for raisins. There may be other fruits and vegetables tainted by wildfire smoke.

Hard to Think

We know in a broad, general sense the shape of the future that warming temperatures and higher CO_2 levels will bring. We know, for example, that sea level rise will continue until an equilibrium is reached between atmospheric

CO_2 levels and melting ice. We know that plants and animals will move to cooler places. We know parts of the Earth are becoming uninhabitable, at least for people living and working outside. Yet as the future draws ever more sharply into focus, new details will emerge, details that will surprise, perplex, and dismay. The devil lies in the climate details.

One of the more surprising details to come to light recently is that increased CO_2 concentrations apparently diminish human cognitive abilities. That's because higher CO_2 levels in our blood reduce the amount of oxygen reaching our brains. Researchers at Harvard and Syracuse University demonstrated a 15% decline in human cognition at CO_2 levels of 945 parts per million (ppm)—about twice today's levels and likely close to the atmospheric levels toward the end of this century. At 1,400 ppm, cognitive decline is 50%. The same correlation holds true for air pollution and human cognition. Higher concentrations of ozone and fine particulate matter ($PM_{2.5}$ levels) impair our ability to think. Hotter temperatures also make it hard to think clearly. Try solving a difficult math problem outdoors on a hot and muggy summer afternoon. At a time when we need our collective wits to make wise decisions regarding climate change mitigation and adaptation, our brains may not be up to the task due to hot temperatures and higher CO_2 levels.

What to Do

- **Check the air quality.** You can gauge the quality of the air around you by checking your local air quality index. A useful tool is EPA Smoke Sense, a free app that provides an air quality map and recommendations for what you should or shouldn't do. AirNow.gov provides local air quality alerts via the EnviroFlash email service.
- **Consult your doctor.** If you have heart or lung problems, talk with your doctor before fire season to plan when to leave the area, and review your asthma plan with the doctor.
- **Develop an asthma action plan:**

  Create a list of triggers and how to avoid them.

  Make sure you have your medication and know how to use it.

  Write down (for yourself and others) what to do during an asthma episode.

  Learn when to call a doctor.

  Create a list of emergency phone numbers.

- **Stock up on staples.** Have several days' supply of food that does not need gas cooking, which would add to indoor particle pollution.

Figure 32. The US Environmental Protection Agency (EPA) provides DIY instructions on how to construct your own air filter to improve the quality of air in your home. Image from EPA (2018a).

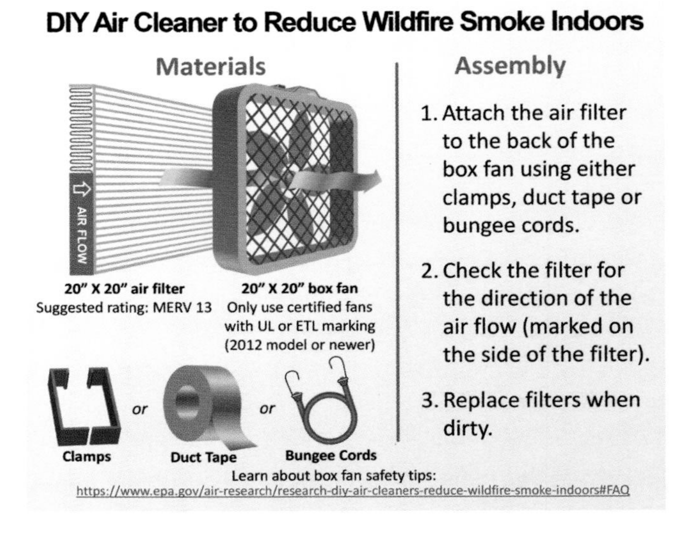

- **Reduce indoor pollution.** Don't smoke cigarettes, and don't use gas, propane, or wood-burning stoves. No vacuuming, no candles, and no incense. Don't fry or broil food.
- **Use efficient air filters** in central air systems.
- **Reduce outdoor physical activity.**
- **Consider purchasing an air cleaner** (EPA 2018a) or make your own (see figure 32).
- **Create a clean room.**
- **Keep kids (and pets) indoors**, and don't involve them in cleanup work.
- **Have a supply of N95 or P100 face masks** ready for outdoor work.
- **Avoid dusty areas**, including construction sites. Pay attention to extreme weather events like dust storms and wildfires.
- **Winegrowers:** There are numerous websites for companies claiming to have developed ways to eliminate smoke taint from grapes. Whether the claims are legit or not is another matter.
- **Research: Particulates and wind.** More research is needed to evaluate the risk of pathogen transport on wind. Will the anticipated uptick in dust storms spread more diseases? Should we be fearful of pathogens piggybacking on particles in wildfire smoke? What about mercury, lead, arsenic, cadmium, and other toxic metals from China's coal plants blown into the waters of the Pacific?
- **Research: CO_2 and cognition.** Scientists need to clarify the link between air pollution and cognition. Most scientists agree there is some diminishing of our decision-making ability in the presence of

air pollution, hot temperatures, and high levels of CO_2, but some recent studies have yielded ambiguous results. If the links between diminished cognition and bad air turn out to be accurate, what can we do about it?

- **Environmental justice.** As a society we must recognize the impact of bad air quality on low-income communities not just in the United States but in poor communities around the world as well. We must do better in addressing these communities' health needs and in adapting the impact of climate on air quality.

FIRE

Wildfires

But clouds bellied out in the sultry heat, the sky cracked open with a crimson gash, spewed flame—and the ancient forest began to smoke. By morning there was a mass of booming, fiery tongues, a hissing, crashing, howling all around, half the sky black with smoke, and the bloodied sun just barely visible.
—Yevgeny Zamyatin, *The Dragon: Fifteen Stories*

The Kingdom of Fire

As the United Nations panel put it in their 2022 report, wildfires are "spreading like wildfire" (UNEP 2022). They are spectacular in the death, destruction, and devastation they cause and the difficulty we have in stopping them or escaping them. Wildfires are rapidly becoming the face of climate change. At a time when humanity desperately needs to reduce carbon emissions, wildfires release massive quantities of carbon dioxide (CO_2), smoke, and other pollutants.

Wildfires are on average getting steadily larger, causing more damage to people's homes and degrading ecosystems. The global increase in temperature is making wildfires more intense, hotter and more widespread. Fire seasons have been extended. Wildfires are no longer confined to forests or grasslands but burn right into our communities. They are much feared now, these crimson dragons that hiss and howl and crush our hopes for a happy life. This is especially so in the western United States (see figure 33).

Figure 33. A 2021 California wildfire that is consuming barely visible buildings. The California fires reflect a warming atmosphere and soils desiccated by years of drought. Photo from NOAA.

The new United Nations Environment Programme report *Spreading like Wildfire—the Rising Threat of Extraordinary Landscape Fires* (UNEP 2022) predicts that especially severe wildfires will increase 14% by 2030, 30% by 2050, and 50% by the end of the century (unless we do something to halt global climate change). The damage from these fires is exacerbated by growing numbers of people moving into what is essentially wilderness, mostly unspoiled, beautiful country perhaps, but the land of John Muir and Edward Abbey is also the kingdom of fire.

What Causes a Wildfire?

Wildfires are uncontrolled fires in forest, grassland, or peat bogs (aka peatlands). The degree of danger from a wildfire is a function of heat and wind, aided considerably by the dry conditions created by droughts, higher temperatures, and increasing rates of evaporation. Topography also plays a role; fire moves more quickly uphill than it does downhill. Factors that influence fire behavior are shown in figure 34.

According to the National Park Service (2022), people start 85% to 90% of all wildfires in the contiguous United States, either by accident or in deliberate acts of arson. However, in Alaska and many national parks, lightning ignites the majority of wildfires, though occasionally fires are triggered by

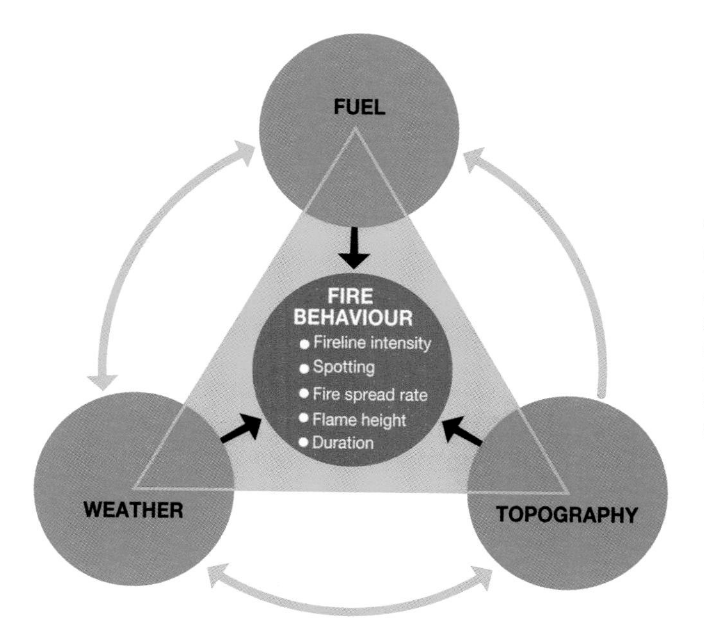

Figure 34. Fuel, weather, and topography determine a fire's intensity, size, height, and duration. *Spotting* refers to the spread of sparks or embers. Topography plays a role because fires move faster uphill than downhill. Image by GRID-Arendal / Studio Atlanti, a UNEP partner (2021); https://www.grida.no/resources/15551.

volcanic eruptions and also, as bizarre as it sounds, by meteorites. For a fire to begin, no matter what the source, three things are needed: heat, fuel, and oxygen. Important causes of wildfires are:

Burning debris. Sparks from burning plants, scrap lumber, and the like can travel for miles. This is particularly true on windy days. A related situation is the escape of a controlled burn, a fire started by Forest Service or Park Service individuals to clear out brush.

Unattended campfires. These are one of the leading causes of wildfires. The 2007 Ham Lake Fire in Minnesota was started by a campfire, eventually burning 75,000 acres (~30,000 ha).

Electrical power. Fallen power lines create 10% of California's wildfires (400 fires per year), including the 2018 Camp Fire that destroyed the town of Paradise (Atkinson 2018; Ganey 2019; Mohler 2019).

People. For whatever dark psychological reasons, something like 21% of wildfires started by people are by arsonists (Daley 2017). Other ways people start wildfires include discarded cigarettes, car crashes, fireworks, and various activities involving matches.

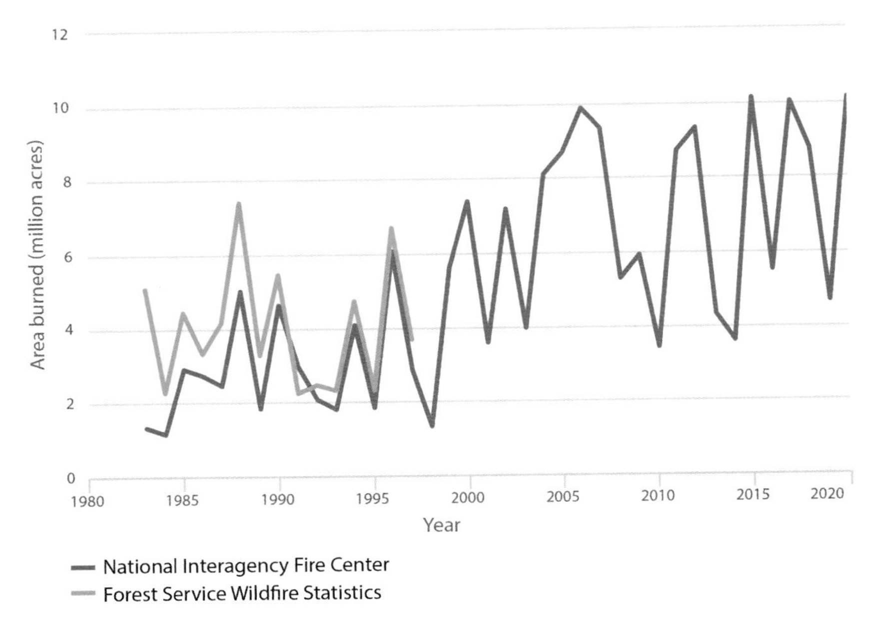

Figure 35. The figure illustrates the annual area burned by wildfires in the contiguous United States from 1983 to 2020. The two lines represent two different reporting systems. The Forest Service (orange line) stopped collecting statistics in 1997. Although there are annual variations, the area burned by wildfires is clearly increasing. Image from the US Forest Service and the National Interagency Fire Center.

Big Burns

The largest wildfire in US history was probably the Big Burn Fire in 1910, which consumed 3 million acres (1.2 million ha) in eastern Washington, Idaho, Montana, and southeastern British Columbia. There were fires in the 19th century of a similar scale, prompting skeptics to declare climate change and its influence on wildfires to be a hoax. But firefighters were fewer in those days, their equipment primitive by today's standards, and cooperation between firefighters from different countries or states nonexistent. Forests, at least in some regions, were larger than today and had fewer roads offering access to fires. Firefighters parachuting into wilderness to fight forest fires (smoke jumpers) didn't get started until the 1940s. Back in the day, the big fires raged largely unchallenged.

In the United States, the area burned each year by wildfires has increased since 1980 (see figure 35). According to NASA satellite photos, however, the total area of land burned globally each year declined by 24% from 1983 to 2015. This is due to changing land use in the tropics, where a shift from nomadic lifestyles to sedentary agriculture (especially in Africa) has converted

Figure 36. State outline maps showing acres burned in 2021, illustrating how vulnerable the western US states are to fire. Image from Move.org.

grasslands and savannahs that used to burn regularly to farms, ranches, and communities, where wildfires tend to be better controlled. No farmer wants to see their fields burned.

There are still severe wildfire seasons that burn large tracts of land. In the 2004 Alaskan wildfire season, 6.6 million acres (2.7 million ha) of land burned. Texas had a particularly bad fire season in 2011. Drought, strong winds, and high temperatures conspired to unleash over 31,000 fires that scorched 4 million acres (1.6 million ha) and destroyed almost 3,000 homes. Historically, severe wildfire seasons also occurred in the upper Midwest, but these days the lion's share of American megafires are in the West (see figure 36).

Canada is no stranger to large wildfires. In 2023 the country went through the worst fire season in the nation's history. All provinces and territories, from British Columbia to Nova Scotia, were stricken by innumerable wildfires. By late summer, an area larger than the state of North Carolina, more than 34 million acres (13.8 million ha) had burned. Smoke from Ontario's fires tainted skies as far south as Florida, prompting air quality warnings for Americans to stay indoors. For a while, New York City had the most polluted air on the planet. The *Washington Post* even jested that humanity had entered a new geological epoch . . . the Pyrocene!

The Yellowstone Lesson

There is much to be learned by studying the history of major fires. A good example is the 1988 fire in Yellowstone National Park. When the fire broke out, officials in the US Forest Service (backed by scientists) declared a new policy for firefighting. The policy was essentially to do nothing, which of course is Nature's way. Before the arrival of settlers, the forests of the western United States were beautiful, yet no one was putting out the fires. So why shouldn't we do what Nature has been doing for thousands of years? In fact, Native Americans for millennia deliberately started wildfires to clear out brush and improve hunting conditions. So when the 1988 Yellowstone fire started, firefighters were told to stand by.

Strong winds and ongoing drought conspired to spread a scattering of small fires into an inferno. The fire burned for months, eventually destroying almost 800,000 acres (324,000 ha) of forest. It was a catastrophe for the park, for park visitors, and of course for wildlife.

The no-fight fire policy was canceled, and the Yellowstone ecosystem is slowly returning to its former pristine glory. When we visited the park on a family trip in 2014, blackened tree stumps were still visible in parts of the forest, though the land was clearly on the mend and the wildlife still abundant.

The Deadliest Wildfire in California's History

The fire started on Thursday, November 8, 2018, when an aging power line pole fell to the ground. The hot electrical lines set fire to local bushes and trees, and then . . . away it went. The Camp Fire, as it came to be known, named after its starting point on Camp Creek Road in Northern California, killed 85 local residents and five firefighters and injured 112 people in a matter of hours (Cal Fire 2023; Mohler 2019).

Drought was a significant factor behind the catastrophe. The town of Paradise normally receives 5 inches (12.7 cm) of rain per year by November. But by November 2018 only 1/7 inch (~0.35 cm) of rain had fallen for the entire year. Dry winds in the weeks prior to the fire added to the severity of the drought. The fire was finally contained on November 25, when the first winter rainstorm of the season occurred. Ironically, a red flag warning (warning of the possibility of a wildfire) had been issued on the day of the fire because of gusty winds. The winds apparently increased significantly during the fire.

Most of the damage caused by the Camp Fire occurred within the first four hours. The fire burned more than 150,000 acres (~61, 000 ha) and destroyed more than 18,000 structures in the area. The towns of Paradise (population 4,764) and Concow (population 402) were essentially completely destroyed, each losing more than 95% of their structures.

Traffic jams occurred on evacuation routes out of Paradise, which led to abandonment of cars with the passengers continuing their escape on foot. Some of the abandoned cars were bulldozed off to the side to clear the escape route. Patients in the Paradise hospital were evacuated in the private cars of nurses, maintenance workers, and doctors just before the hospital was destroyed. Many residents survived by seeking refuge in large buildings, lying down in commercial parking lots, or jumping into a reservoir. Seventy of the 85 fatalities occurred among those who stayed in town, and the other 15 happened during evacuation—good evidence favoring evacuation from catastrophes when possible (Griffin 2021).

Smoke Jumper

I was a Forest Service firefighter for five fire seasons while a college student, three of those seasons as a Missoula, Montana, smoke jumper. Smoke jumpers were employed to get to fires quickly, and because the jump spots were very small clearings, we used steerable parachutes.

In my two seasons with ground-based fire crews, we carried heavy packs along with tools, often hiking to fires on poorly maintained trails and sometimes on no trails at all. Upon arrival we were often exhausted. Yet strangely, we usually felt fresh, ready, and raring to get to work.

Putting out fires was much different then because the fires were much different. The largest fire I was involved with (which happened to be my first fire) was a 2,000-acre (~800 ha) burn in northern Oregon. Local jails were emptied to make up fire crews. More commonly the fires were an acre or two, many the size of a living room. We all knew about the 1910 Big Burn Fire. But nothing of the sort occurred for the next 110 years until global warming kicked in and dried out the forests with extreme heat waves and droughts.

We fought those small fires back then by clearing a 2-foot-wide (0.6 m) path of forest floor around the creeping flames. We never used water since water pumps and fire hoses were not available in the wilderness. Today on TV, we see firefighters hosing down big fires using water from tanker trucks whenever a fire happens to be near a road. We never had it that easy.

But things have changed. These days, fires that can leap over interstate highways cannot be halted with a 2-foot-wide (0.6 m) cleared path. Now an airborne shower from an air tanker packed with fire retardant is a more effective tool to fight fires in wilderness areas. But of course the most effective firefighting tool of all is natural rainfall. This was also true back in the day. A total of 78 firefighters died fighting the Big Burn Fire, but the fire wasn't contained until a rainy cold snap moved in.

Another difference in wildfires today is that the fire season extends well into mid-to-late fall. However, I was able to go back to college in early September after every fire season between 1951 and 1955.

— Orrin Pilkey

The Mighty Sequoias of California

General Sherman, a 275-foot-tall (84 m), 2,500-plus-year-old sequoia tree, is the world's biggest tree by volume. It has survived many fires over the years. In 2021, as the KNP Complex Fire (a fire caused by lightning) burned nearby, firefighters wrapped the base of the General Sherman (and other sequoias) in aluminum blankets, a technique often used to save humans and buildings from fire. The wrapping is meant to prevent embers from getting into the trees through old tree scars (known as *catfaces*) that run vertically up and down the trunks (Chappell 2021). Tree wrapping has become a standard fire prevention practice in recent years. The wrapping saved the General Sherman, but thousands of other sequoias succumbed to the flames.

How Fares the Rest of the World?

Europe From Ireland to western Russia, from Scandinavia to the Mediterranean, Europe is experiencing its share of wildfires fueled by intense heat waves. In 2018 Sweden suffered 41 times its 10-year annual average of acreage burned by wildfires. Turkey, bombarded by heavy winds, battled more than 100 blazes in 2021, which forced hundreds of evacuations by sea from coastal communities. In both 2021 and 2022, fires broke out near ancient Olympia, burning homes and suffocating the city of Athens with so much smoke that citizens were ordered to stay inside (Porterfield 2022).

Europe's record-breaking 2022 heat wave spawned fires that nearly quadrupled the average annual acreage burned. More than a million acres (405,000 ha) of forest in Greece, France, Spain, Italy, Portugal, Romania, Germany, and parts of Scandinavia were destroyed (Henley 2022). Wildfires once confined mostly to lands near the Mediterranean have expanded north into central and northern Europe. Sadly, fires have ravaged some of Europe's most precious national parks, including the Serra da Estrela National Park in Portugal.

Russia. Siberia has been warming 2.5 times faster than the rest of the planet. As a result, Siberian fires, mostly in permafrost soil, are huge. In 2021 the area burned by wildfires in Siberia (about 19 million acres, or 8 million ha) was greater than all the other fires of the world combined. The black, smoky skies of Siberia reached Moscow and caught the attention of Vladimir Putin, who ordered more efficient efforts in halting the fires. One reason they are now so large is the shortage of firefighters. Normally soldiers make up a large part of fire crews. But the army is on duty in faraway Ukraine, while the forests of Siberia burn largely unattended.

Adding to Russia's problems containing ever-expanding burns in the wilds of Siberia are the so-called zombie fires. These are fires that die in winter only to reignite in the spring and summer. They occur in Siberia because its soils are often made of peat, partly decomposed plant matter that can burn underground for months and even years. In 2020 Siberian wildfires (zombie or otherwise) emitted CO_2 emissions that were roughly the same as the annual emissions of Canada. This represents about 1% of the world's total annual carbon emissions. Zombie fires are troubling for two reasons: (1) like zombies everywhere, they are hard to kill (hard to put out), and (2) they are densely packed with carbon (Sevrin 2022).

South America. The countries in South America, especially in the Amazon rainforest and the Pantanal (where Paraguay and Bolivia border Brazil) have seen a surge in wildfires over the past several decades, partly from climate change and partly from burns to clear land for farms and ranches. Figure 37 shows active fires (in red) on August 24, 2006, in the Pantanal region, and in portions of Argentina. This amazing photo from NASA's Aqua satellite illustrates the extent of wildfires in that part of the world.

Chile has been afflicted with drought since 2010 (see "Drought" in "Water" for details), the worst drought in a thousand years. An intense climate-change-induced heat wave, parched vegetation, strong summer winds (normal for Chile), and accidental fires started by people conspired to ignite hundreds of fires across the country in the summer of 2023 (February in Chile). More than a million acres (405,000 ha) burned.

Patagonia, in southern Argentina, is one of the least developed and most beautiful places left in the temperate regions of the Americas, with crystal-clear lakes and forested mountains, much like the Cascades or the Canadian Rockies. Historically, large wildfires in the region were relatively rare, but cli-

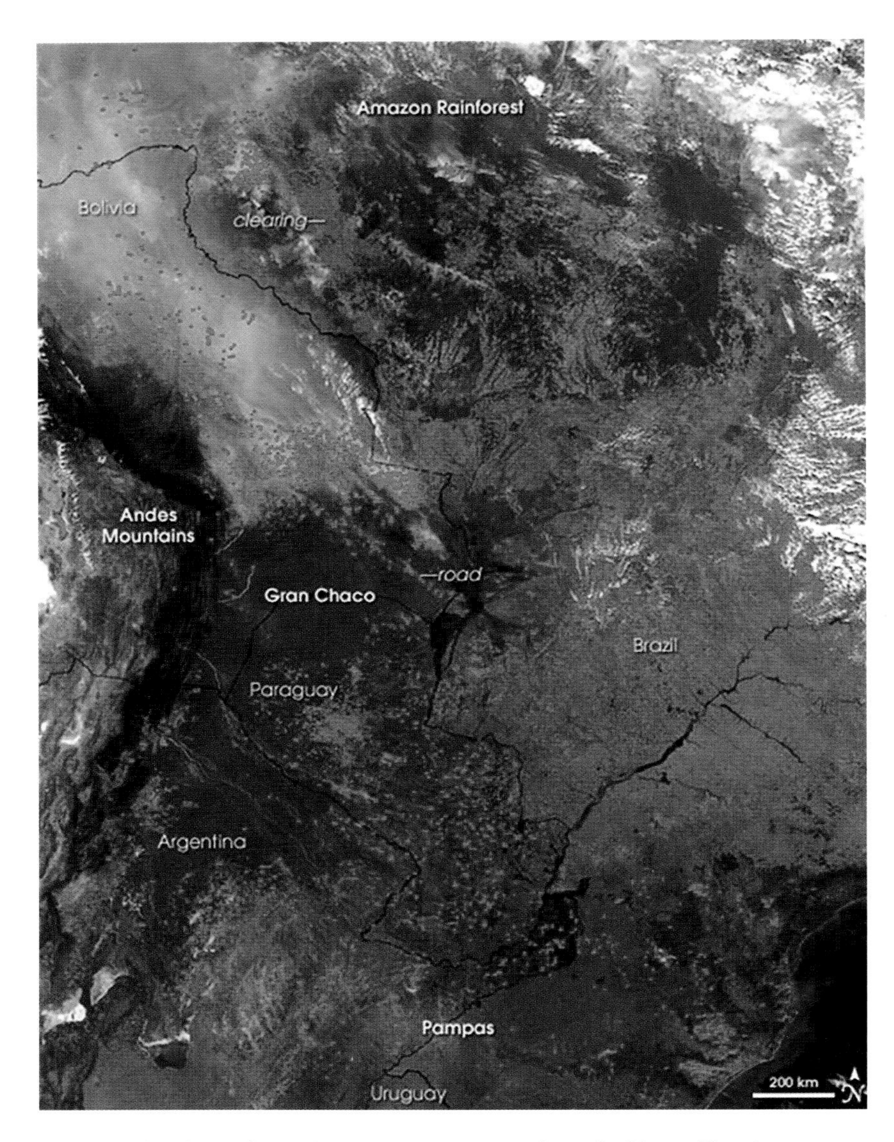

Figure 37. Fires in South America on August 24, 2006, shown in this satellite photo from NASA's Aqua satellite. Probably most of these fires were set illegally by people to burn forests to make farms and ranches. Image by Jeff Schmalty, Modis Response Team, NASA.

mate change and human intervention with the wet forest ecosystem in Patagonia have increased the number and magnitude of wildfires. Patagonian wildfires follow the same causal pattern as those in North America. Diminishing rainfall and a warming climate increase the chances of a fire getting started by lightning, human negligence, or human malevolence. Thomas Kitzberger, professor of ecology at the National University of Comahue in Bariloche, Argentina, points out that settlers in the area for decades planted extremely flammable pine trees imported from North America on their plantations. The combination of volatile, non-native trees plus a warmer, drier climate is fueling the wildfires (Kitzberger 2022).

Brazil has been devastated by wildfires for several years now, because of an amalgam of climate change and human activity. The human involvement is largely in the Amazon rainforest, which is fast being deforested to make space for farms and cattle ranches. Former President Jair Bolsonaro favored policies that continued the destruction of the Amazon rainforest, in spite of his claims to the contrary. Upon his reelection in 2022, President Luiz Inácio Lula da Silva promised to eliminate deforestation in the Amazon, a welcome shift in attitude from the perspective of climate mitigation. We shall see if he's as good as his word.

Preserving the Amazon's rainforest is critical. About 10% of the world's land species live there. Rainforest cleared by burning for agriculture reduces biodiversity and is a huge loss for the Indigenous minority living there. Scientists fear that if deforestation continues in Amazonia, mass extinctions will follow and the rainforest will no longer function as a carbon sink but will become a net emitter of CO_2.

Amazonia garners all the attention when it comes to Brazilian wildfires. But fire impacts other regions of Brazil as well. The Pantanal is the world's largest wetland. Located in Brazil's southwestern states of Mato Grosso and Mato Grosso do Sul and extending into Paraguay and Bolivia, the Pantanal has seen fires four times the size of the largest fires in the Amazon. Temperatures exceeding 105°F (40.6°C) and winds sometimes reaching 80 mph (129 kph) exacerbate the situation. Farmers there report that the fires and high winds are removing much of the topsoil.

Australia has *bushfires* rather than wildfires. It is unfortunate that the native trees in Australia (including eucalyptus trees) are so flammable they can literally explode in a fire, making widespread bushfires unavoidable. Eastern Australia is considered one of the most fire-prone regions in the world due to a combination of flammability, a warming, drying climate, and high winds.

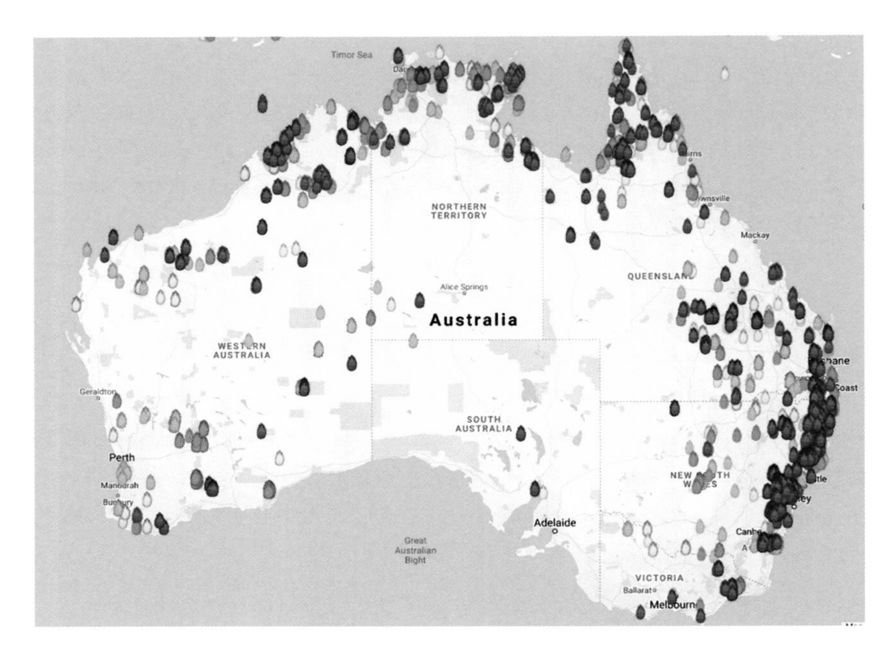

Figure 38. Location of bushfires in Australia in 2019–20. Dark red dots mark the most serious and long-lasting fires in that time frame. Yellow dots are locations of past fires. The fires have taken a huge toll on Australian wildlife. Image from the Center for Disaster Philanthropy.

As in the United States and other parts of the world, Australia's fire season is getting longer.

In the "Black Summer" of 2019–20, 15,000 wildfires burned at least 47 million acres (19 million ha) of bush, killing more than a billion animals. The 2021 documentary film *Burning,* directed by Eva Orner, depicts in grim detail the apocalyptic horror of Australia's Black Summer wildfires and the inability of its government to come to grips with a changing climate. The film is a sobering reminder of the severity of climate change made worse by the intransigence of politicians. *Burning* may well be a sneak preview of more horrors to come (Cart 2022). (See figure 38.)

But the fires of Black Summer may portend more than a future of ravaged ecosystems and increased carbon emissions. Smoke from wildfires apparently can damage the ozone layer, the same ozone in the upper atmosphere that blocks harmful solar radiation. Scientists are still puzzling over the chemical interaction between smoke and the Earth's ozone layer, but it is widely believed Australia's recent wildfires significantly widened the hole in the ozone high over Antarctica (Coleman 2022).

What to Do

- **Discourage development** (especially residential) near fire-prone forests through smart zoning rules. Building houses in a fire-prone "wilderness" is as foolhardy as building houses on a floodplain or on an eroding beach.
- **Increase the space** between structures and nearby trees and brush, and clear space between neighboring houses.
- **Incorporate fire-resistant design** features and materials in buildings. Houses built of brick, concrete, steel, and stone are better at resisting fires than wood. Underground houses and adobe houses are also fire resistant (no structure is 100% fireproof).
- **Increase resources** allocated to fire prevention, including more aluminum blankets to wrap important trees threatened by fire.
- **Remove fuels** such as dead trees from forests that are at risk. The natural forest floor should be cleared away from a building's edges.
- **Develop recovery plans** before a fire hits, and implement plans quickly after a fire to minimize habitat damage, reduce potential erosion, and prevent flooding and mudslides on denuded slopes.
- **Check the Environmental Protection Agency (EPA) website AirNow.gov** to view interactive air quality maps, including a map of current wildfires in the United States.
- **Do controlled burns.** Small controlled fires can help perpetuate forest ecosystems by eliminating undergrowth and dead vegetation that often fuels larger fires. Some trees, like jack pines (*Pinus banksiana*) and sequoias, require a fast-moving fire for propagation, as heat causes their cones to open and release the seeds.
- **Learn from Indigenous people.** Native Americans and First Nations people in Canada have developed techniques over the centuries to keep catastrophic wildfires at bay through controlled burning.
- **Conduct fire drills.** Practice fire evacuation drills with your family and with pets. Make a bug-out bag, including respirators (see "Bug-Out Bags"). Make an evacuation plan, including mapping escape routes. Keep your gas tank full during fire season. Wear long pants and shoes or boots. What will you do if the roads are impassible?
- **Remember your pets.** Have pet containers, pet food, water, pet bowls, and leashes stored in your car to facilitate a quick evacuation. Make sure your pets have identity collars and microchips.
- **Don't wait to evacuate.**
- **Relocate.** At some point homeowners living in wildfire-prone areas will have to make a decision, the same decision those living in coastal

areas affected by rising seas are making. When is it time to move to a safer place?

- **Shift budgets.** Currently, fighting wildfires receives over half of related expenditures, while planning and research receive less than 1%. Changing to a "Fire Ready Formula" involves devoting two-thirds of spending to planning, prevention, preparedness, and recovery, with one-third left for response.

Urban Firestorms

It wasn't a wildfire in the forest, it was a suburban and urban fire. The Costco we all shop at, the Target we buy our kids' clothes at—all damaged.—Jared Polis, governor of Colorado

The Marshall Fire

On December 30, 2021, people exiting stores were startled to see a wildfire just down the road . . . a fire that hadn't been there moments before. The fire had come from nowhere. At least that's the way the citizens in a densely populated Denver, Colorado, suburb described it. There were no towering clouds of smoke billowing from a forest. No hellish glow on the tree-lined horizon. In fact, there was no forest at all . . . just flaming grasslands. But most important, there were no loudspeakers warning people to escape and no detailed and publicized getaway plans.

The small towns of Louisville and Superior, 9 miles (14.5 km) from Boulder and part of Denver's suburban sprawl, were completely devastated by the Marshall Fire. What were once typical middle-class American neighborhoods of manicured lawns and sidewalks transformed in a few hours into a melted, misshapen ruin. But this was a different kind of fire from Colorado's 2020 Cameron Peak Fire that laid waste to 200,000 acres (81,000 ha) of Col-

Figure 39. A Boulder County neighborhood outside Denver, Colorado, after the Marshall Fire roared through. The fire began in nearby grasslands and was fueled by high winds. A total of 1,084 houses were destroyed, in large part because they were built too close, allowing flames to jump from house to house with ease. Credit: Hart Van Denburg / Colorado Public Radio / Pool.

orado forest. The Marshall Fire was a mere 6,000 acres (2,400 ha) in size, and the acreage burned was entirely urban.

The final toll of buildings in the Marshall Fire was 1,084 structures destroyed and 149 damaged, totaling more than $500 million in damage (see figure 39). This included a Super Target store whose roof proved to be flammable and unable to resist flying embers. A nearby Costco and other businesses in a shopping center were either destroyed or damaged. Around 35,000 people were forced to flee their homes. Miraculously, only two people died in the fire.

The Marshall Fire started in two places: sparks from a power line and a fire on residential land that was thought to have been put out but reignited a few days later (McKinley and Pearce 2023). Once started, the two fires merged and quickly spread. They were aided by winds of more than 100 mph (161 kph) that burned grass, which that year was unusually dense and tall from higher-than-average spring rainfall. The rains had been followed by months of severe drought (the driest in 150 years). Unseasonably warm temperatures, little snowpack to supply water to dampen vegetation, strong winds . . . perfect conditions to start a wildfire. Fortunately, New Year's Day delivered a light snowfall, providing just enough snow to squelch the still-smoking embers.

Experts said that four problems contributed to the magnitude of the disaster: (1) high winds that blew for eight straight hours; (2) the proximity of the community to flammable grasslands; (3) the tight spacing of the houses, often with 10 feet (3.1 m) or less between them; and (4) too much mulch added by residents to help their yard plants endure the drought.

Carole Walker, director of the Rocky Mountain Insurance Information Association, notes that Colorado and the West are locked in a cycle of escalating catastrophes. In response to this crisis, Walker noted that there is "a renewed Colorado push for 'hardening' now in suburbs as well as mountain forest developments" (Finley 2022). *Hardening* means rebuilding communities to make them fire safe. It is an approach that will collide with the developers and planners pushing for higher-density, mixed-use communities. Ironically the term *hardening* is also used to describe the building of seawalls on ocean shorelines to halt erosion and protect buildings.

Tomorrow's Firestorms

A firestorm is an intense and rapid conflagration that creates its own local weather conditions. Firestorms are historically associated either with war, as, for example, with the bombing of Japanese and German cities in World War II, or with extremely intense wildfires. The Marshall Fire represents a new species of firestorm that ignites near cities and destroys buildings. The 2018 Camp Fire that devastated the town of Paradise, California, could also be characterized as an urban firestorm.

The most recent and most tragic urban firestorm was the 2023 fire on Maui in Hawaii—the deadliest US wildfire in over a century. Fueled by dry conditions and near–hurricane force winds, the fires destroyed the town of Lahaina, leaving charred ruins. No warning siren had been sounded and at least 100 people (including children) perished. Many residents were caught unaware, still sleeping in bed. As the climate warms and urban sprawl continues to expand into fire-prone ecosystems, it seems likely that urban firestorms like the one in Lahaina will increase in scale and frequency, causing more death and destruction.

What to Do

- **Increase spacing between buildings.** One way to reduce a fire's domino effect is to space buildings 20 feet (6 m) apart or more. Some homeowners while rebuilding a burned house purchase adjacent properties to give themselves more fire clearance.

- **Design an early warning system for fires.** Cities vulnerable to wildfires should have a siren-based warning system to alert citizens to approaching wildfires.
- **Maintain powerlines.** Sparks from aged or downed power lines can cause wildfires, especially if adjacent vegetation hasn't been cleared. When conditions are dry and windy, power companies might consider temporarily turning off the power to prevent arcing.
- **Plant rocks instead of grass.** A watered lawn may resist burning but a yard made of stone can't burn. Many desert communities are encouraging people to landscape their yard with local plants interspersed with stone or sand.
- **Practice controlled burning.** The widespread practice of fire suppression in forests has increased the density of burnable vegetation and the overall fire risk.
- **Use buffers.** Buffers that resist burning must be required, should be at least 5 feet (1.5 m) wide, and must be kept bare.
- **Remove flammables near your home.** One of the recent findings by wildfire researchers is that urban fires do not spread by flames on the ground but by sparks borne on the wind, spreading in the air overhead from house to house. Eliminate anything around or on your house that might catch fire. This includes mulch, plants, leaves in gutters, outdoor furniture made of wood or fabric, wooden doors, wooden shutters, and the like. You may need to trim trees or replant them away from the house.
- **Update building codes.** The new green (fire-resistant) building code instituted by the town of Louisville, Colorado, will raise the cost of construction but will be worth it in the long run. Building codes should require new houses to be made of fire-resistant materials. Examples include brick, stone, concrete, cement-board siding, fire-resistant glass, and ceramic tile or metal roofing.
- **Plan development with fire in mind.** City planners may want to reconsider allowing development to spread into ecosystems known to have evolved to coexist with wildfires.
- **Buy a goat.** Goats will eat almost anything, including the burnable undergrowth in the woods near your house. In several parts of the world goats are employed to create fire buffers by eating away flammable vegetation. We recommend keeping your goat away from the garden.

WATER

Sea Level Rise

The sea has never been friendly to man. At most it has been the accomplice of human restlessness.—Joseph Conrad, *The Mirror of the Sea*

Here Comes the Sea

Of all the devilries spawned by a changing climate, sea level rise is the most insidious. A hurricane might pound a coast for a day before it dissipates and moves on. A wildfire might burn through a forest in a month until it fizzles out and is soon forgotten. But sea level rise is different. It is neither episodic nor seasonal. The ocean doesn't rise one year and not the next. It just keeps coming, inexorably, relentlessly . . . without pity. Year after year, century after century, the sea will keep rising. And woe be to any man or beast that gets in its way.

This is why sea level rise will likely be the event that finally delivers society its long-awaited moment of clarity, the salty slap in the face of politicians, knocking all pretense and obfuscation from their minds, stranding them in the rising waters of an appalling realization: it's hard to ignore the reality of a rising sea permanently engulfing entire cities. Sea level rise is a different kind of devil altogether.

The Changing Level of the Sea

The only constancy about sea level is its inconstancy. It is forever changing, most notably from the daily tides that respond to the combined gravitational pull of sun and moon. The difference between high tide and low tide can exceed 50 feet (15 m), as in Canada's Bay of Fundy. But in most locales the tidal range is well under 10 feet (3 m). Depending on the orbits of the Earth and its moon, the high tides are highest (and low tides lowest) during the so-called spring tides when the moon is full or during a new moon. Today spring tides reach further inland than before because the sea has risen globally about 8 inches (20 cm) over the last 100 years.

The highest of all spring tides are known colloquially as *king tides*. These occur two to three times per year when the moon and sun are closest to the Earth. They are entirely predictable, and their expected arrival date, time, and magnitude are important items in local newspapers. In most places king tides are only a few inches higher than the average high tide. But when accompanied by river flooding, ocean currents, low atmospheric pressure, or strong onshore winds (winds blowing from sea to land), king tides in the United States can reach more than a foot (30 cm) higher than an average high tide (NOAA 2023). This can cause serious coastal flooding (Sweet et al. 2022). Many have noted that flooding during king tides is a preview of the future, a lesson in how a rising sea will affect those who have chosen to live on the coast (see figure 40).

Minor coastal flooding from tides and wind is called *nuisance flooding*, *tidal flooding*, or *sunny-day flooding*. None of these is a scientific term. A king tide, in effect, is an extreme case of nuisance flooding. As mentioned above, the heights of tides are affected by wind, local rainfall, air pressure, and storms. Tidal ranges are also influenced by the presence or absence of El Niño. Because roads and other infrastructure in coastal cities were built when sea level was considerably lower, nuisance flooding is becoming more frequent and more damaging as the ocean continues rising.

Norfolk, Virginia, on the western shore of Chesapeake Bay, has a much-publicized king tide, perhaps because of its impact on the naval base there and also because the lower Chesapeake area is sinking from groundwater depletion, making local sea level rise proportionately higher. On the anticipated day of a coming king tide, cars are moved to higher parking spots in town, and arrangements are made for the temporary use of alternative parking lots for businesses and churches. A Unitarian church used to publish a tide table in each Sunday program so church goers could safely navigate tidal flooding when driving to and from the church (see figure 41). Eventually, the church building was sold, and the congregation moved to higher ground. Smart.

Figure 40. A five-bedroom, $300,000-plus beachfront cottage was a January 2022 victim of sea level rise and shoreline retreat near Rodanthe, on North Carolina's Outer Banks. Debris from the house was scattered for hundreds of yards down the beach—a long-term hazard for swimmers and walkers. Photo courtesy of the Cape Hatteras National Seashore / National Park Service.

Tsunamis and storm surges are other examples of rapid, temporary sea level changes that are potentially big enough to be damaging. Tsunamis affecting an entire ocean basin are caused by shifting tectonic plates. Local tsunamis may happen when a cliff collapses into a body of water. (See "Tsunamis" for more.)

On the US East and Gulf Coasts, hurricanes are the most common cause of the sometimes sizable inundation by ocean water known as a *storm surge*. In New England and the Canadian Maritimes, extratropical storms (*nor'easters*) are far more important mechanisms for creating storm surges (see "Hurricanes" in "Air"). In general, wider and flatter continental shelves allow larger storm surges. The gently sloping shelf off the Mississippi Delta was responsible for the horrific 30-foot (9 m) storm surge of Hurricane Katrina in August 2005. Storm surges are much smaller off the Outer Banks of North Carolina because the continental shelf there is narrower and steeper.

The Rising Sea: Consequences

One of the truly major global consequences of climate change is sea level rise. In its 2021 bombshell climate report, the United Nations (UN) Intergovernmental Panel on Climate Change (IPCC) noted that accelerating sea level rise is happening now and will continue for decades to come. The report, authored by dozens of scientists, extensively documents their conclu-

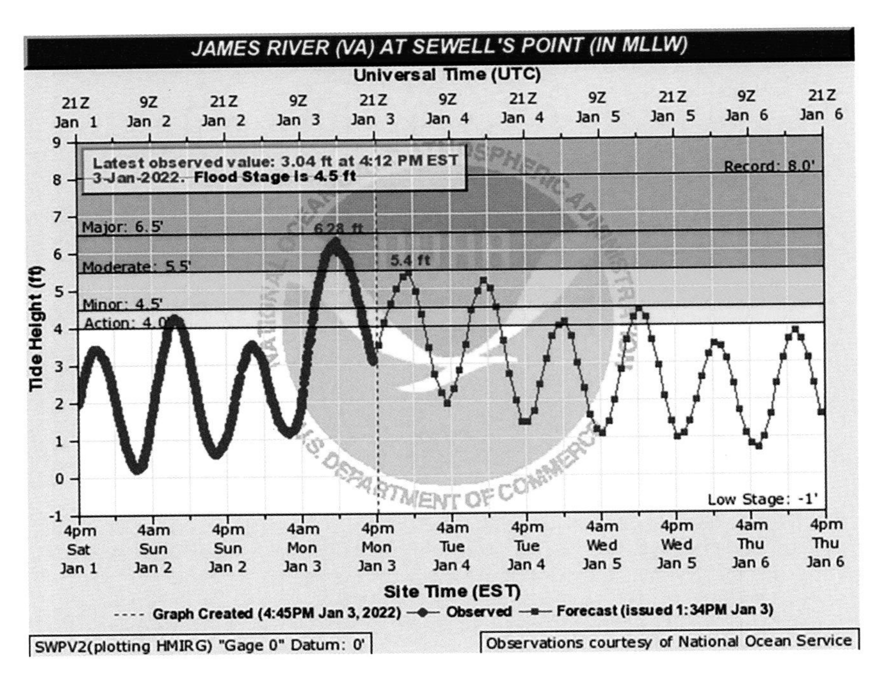

Figure 41. This is an example of the January 3, 2021, warning given to the public in Norfolk, Virginia, about a forthcoming king tide expected to be 1.4 feet (40 cm) higher than an average high tide. Image from NOAA, published and distributed to local citizens by Skip Stiles of the organization Wetlands Watch.

sions and expresses a level of certainty not present in previous reports. The UN report emphasizes the critical impact sea level rise will have on coastal cities worldwide in terms of economic loss, fatalities, and forced migration. According to a Cornell University study, the world will see 2 billion refugees by 2100 just from sea level rise alone (Cornell University 2017). Ironically, as the portion of the planet covered by ocean increases, the total land area available for refugees will diminish.

It is important to understand that sea level rise is not the same everywhere. It varies locally due to land sinking or rising, ocean currents, winds, the gravitational pull of glaciers and mountains, and proximity to large river mouths. A common cause of rapid (local) sea level rise is the extraction of drinking water from groundwater causing the land to sink. This is happening in the lower Chesapeake Bay area. Some regions like the western coast of Canada and parts of coastal Alaska will not experience sea level rise in the immediate future since the lands there are still rebounding (rising) after the ice sheets melted at the end of the Pleistocene.

Unlike disasters such as hurricanes and wildfires, sea level rise per se will generally not involve an immediate need to escape. It will, however, magnify other climate-related events such as the magnitude of storm surges, the height of king tides, the frequency of nuisance flooding, and the rates of shoreline erosion. Millions of people living on low-lying Pacific islands such as the Maldives and the Marshall and Solomon Islands are in particular trouble. Many islanders have already relocated to other countries after rising ocean waters flooded their homes and tainted their crops and groundwater with salt.

Rising seas will cause the partial abandonment of a number of coastal cities. This is already happening in Jakarta, Indonesia, where the capital is being moved to another island. Most of Miami, some other cities in Florida, and much of New Orleans will likely be abandoned this century. Sea level rise will also force the moving and probable abandonment of seaports, oil refineries, power plants (including nuclear ones), sewage disposal plants, parks, hotels and other tourist facilities, and miles and miles of beach cottages and highways. Globally this will trigger a massive climate refugee crisis.

Two mechanisms cause sea level rise: (1) the melting of the Earth's polar ice sheets and (2) the thermal expansion of ocean water. Both result from global warming, caused by humanity's output of industrial greenhouse gases. Because the ocean/air climate system has a certain inherent inertia, sea level rise will continue for centuries and likely for thousands of years.

If Greenland's ice sheets melt, then sea level will rise more than 20 feet (6 m) but only in the Southern Hemisphere. As counterintuitive as it seems, the sea level will actually drop in the northernmost North Atlantic. That's because the gravitational pull of Greenland's massive ice sheets is strong enough to significantly pull the adjacent waters up. As the ice melts, its gravitational pull diminishes, and sea level goes down in the region. Scotland, Norway, Iceland, parts of Canada, and Greenland will increase in land area. Greenland will gain even more land in the coming centuries as it rebounds from the weight of all the ice that formerly pressed down on the island.

The same is true for Antarctica. Once the southern ice sheets melt, sea level will sink around the South Pole and rise in the North Atlantic. If all the planet's ice melts, including the Greenland and Antarctic ice sheets, then the world's oceans will rise everywhere about 230 feet (70 m).

It should be noted that the melting of the planet's sea ice will not raise ocean levels. But it will indirectly affect global sea levels by allowing land-based (mostly mountain) glaciers to slide more quickly into the sea. The sea ice (especially in Antarctica) functions as a kind of dam holding back continental ice sheets. Once that sea ice melts . . . look out!

Oceanographer John Englander found that in the year 2000 the sea was rising 19 inches (48 cm) per century, at a rate that had doubled in just 20

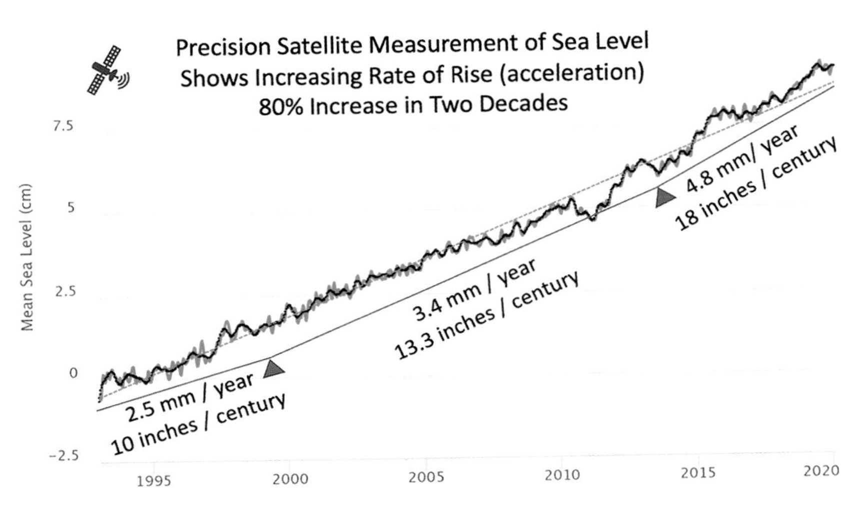

Figure 42. Generalized sea level curve for the quarter century from 1995 to 2020. The data are from open-ocean satellite observations and clearly show a significant acceleration (more than doubled between 2000 and 2020) in the rate of sea level rise. The observed rates of ice-sheet melting in both the Arctic and Antarctic indicate that this rise of the sea will continue to accelerate. According to NOAA, sea level rise will be somewhere between 3 feet and as much as 8.2 feet (1 m to 2.5 m) by the end of this century. Courtesy of John Englander.

years. Englander's numbers from the open ocean have the advantage of not being affected by local land movements (see figure 42).

The projection of the National Oceanic and Atmospheric Administration (NOAA) puts the range of sea level rise from a minimum 2 feet (0.6 m) from carbon emissions already in the atmosphere to 3.5 feet–7 feet (1.1 m –2.1 m) if we moderately curb future emissions, and to a maximum of 8.2 feet (2.5 m) by the end of this century if carbon emissions continue rising. Put simply, the greater the emissions, the higher the sea will rise. Extrapolation from the 8.2-foot maximum means a 2-foot (0.6 m) sea level rise by 2046 and a 3-foot (0.9 m) rise by 2059 (Sweet et al. 2022). An ocean that is 2 or 3 feet higher will make most barrier islands in the world uninhabitable (see table 9), according to University of Miami geologist Hal Wanless (2021). You folks living on Hatteras (on the Outer Banks of North Carolina) . . . are you paying attention?

What does a single foot of sea level rise mean on a barrier island? North Carolina affords an extreme example. The average slope of the land immediately behind most of North Carolina's barrier islands is 1:2,000, meaning that in theory, flooding behind the barrier islands could extend back landward 2,000 feet (610 m) in response to a 1-foot (30 cm) rise in sea level. On

Table 9. Barrier Islands around the World

Country	Number of Islands	Shore Length (miles)
The World	2,152	13,035
United States	405	3,054
Russia	226	1,020
Australia	208	995
Canada	154	294
Madagascar	119	377
Mozambique	115	563
Mexico	104	1,392
Brazil	72	559

the sound side of the Outer Banks of North Carolina, the slope of the outer coastal plain is 1:10,000. This means that a sea level rise of 1 foot could cause flooding through inlets for nearly 2 miles (3.2 km).

There are many factors that may affect the rates of future sea level rise. One possible blip in the US sea level curve is the future behavior of the Gulf Stream, which maintains Europe's relatively mild climate by bringing north warm water from equatorial regions. Cold, fresh water flowing from melting Greenland ice threatens to stop the flow of the Gulf Stream, causing the climate of Europe to cool significantly. This halted flow could force the orphaned Gulf Stream to be pushed back toward North America, increasing the rate of sea level rise on the US East Coast.

Some bad news concerning Antarctica came from the December 2021 meeting of the American Geophysical Union in New Orleans. According to Erin Pettit, a glaciologist from Oregon State University, and colleagues, a large floating ice shelf along West Antarctica is showing strong signs of breaking up (Pettit et al. 2021; Benn et al. 2021). The ramifications of this breakup, which could be completed within 5 years, are that it will release the Florida-sized Thwaites Glacier (the world's widest glacier). The melting glacier will subsequently flow rapidly into the ocean, raising ocean waters by as much as 2 feet (60 cm) in the North Atlantic. As the Thwaites Glacier collapses, it may cause other glaciers to break apart, taking most of West Antarctica's ice with it. If this happens, sea level could rise as much as 10 feet (3 m) within a few decades. Similar rapid spurts of sea level rise, known as *meltwater pulses*, have happened in the recent geologic past. In 2022 and early 2023, the rate of ice melting in Greenland, West Antarctica, and the

Figure 43. BEFORE: August 1941 photo of the Muir Glacier in Alaska by W. O. Field, in what is now Glacier Bay National Monument. The glacier at the upper right is the tributary Riggs Glacier. Thousands of melting glaciers like Muir are partially responsible for sea level rise. Photo from USGS, provided by Bruce Molnia.

Antarctic Peninsula increased significantly. This will translate into a slightly accelerated sea level rise.

There is some bad news from Alaska as well. Figure 43 shows the Muir glacier in 1941. The same glacier is shown in 2004 (figure 44). A significant contribution to sea level rise comes from the melting of mountain glaciers like those in Alaska.

If all the above events unfold as predicted, the Thwaites Glacier breakup could contribute a full 25% of the annual sea level rise, compared to today's contribution of 4%, according to the British Antarctic Survey. At the same time, other reports from the 2021 American Geophysical Union meeting show that the Arctic is warming at an unexpectedly high rate, four times faster than the rest of the world.

The extraction of groundwater tends to compact underlying sediment, a common reason why land sinks. An extreme example of this problem is Jakarta, Indonesia, where much of the city is already below sea level and protected by levees and walls. The sea level there is rising (or the land is sinking) as much as 1 foot (30 cm) per year, largely due to groundwater extraction from numerous, often illegal, wells. Not surprisingly, plans are afoot to move the entire capital and much of the surrounding city to another island with more elevation and less compactible soil.

Figure 44. AFTER: October 31, 2004, photograph of the Muir Glacier illustrating the extent of its retreat and apparent massive contribution of meltwater. Only the smaller tributary Riggs Glacier remains in view. Photo © Bruce Molnia.

The Indifferent Sea

Why worry about escaping sea level rise? After all, sea level rise is incrementally slower than a wildfire, a hurricane, or a flooding river. It would seem there's plenty of time, years or even decades, to avoid the ocean's wrath. But as mentioned earlier, sea level rise can speed up other processes. The word of the day is *synergy*.

If you plan to purchase coastal property to eventually pass on to your children or grandchildren, think again. All indications are that sea level rise is accelerating, storms are intensifying, rates of erosion are increasing on most shorelines, and coastal real estate in many low-lying parts of the world will be rendered worthless in the coming decades. Why risk your children's inheritance by investing in a fool's watery paradise?

The impact of the sea level on US coastal property depends greatly on which coast you live. The West Coast, with its long stretches of rocky cliffs where many neighborhoods are safely distanced from the rising ocean (though not safe from landslides caused by shoreline erosion), is dramatically different from the East Coast. The East (not counting New England) and Gulf Coasts, by contrast, consist of a broad, gently sloping coastal plain, fronted by a 3,000-mile-long (4,800 km) chain of barrier islands stretching from Mexico to Long Island.

There are more than 2,100 barrier islands worldwide. They are generally hazardous places to live and will become more so as the sea rises. It's not that they are at imminent risk of submergence. The problem is that low areas on the islands will be flooded and overwashed by storm surges, cutting off evacuation routes with numerous breaks in roads and the formation of new inlets. This is exactly what happened during the November 7, 2021, nor'easter that struck the Outer Banks of North Carolina simultaneously with a king tide. Highway 12, the single road connecting all communities on the Banks, was washed over in a number of places and made impassable by sand deposited on the roadway. No one could get on or off most of the Outer Banks for several days. And that was from a minor storm. In contrast, Hurricane Ian struck Florida's west coast in October 2022 as a Category 4 storm, wrecking development, large and small, on the barrier islands. Communities on Sanibel Island and Fort Myers Beach on Estero Island, plus the causeway or bridges leading to them, were destroyed. Ian is an example of the rapid intensification of storms as a result of climate change.

While the US West Coast has higher beachfront elevations and more irregular and rocky topographic features that should, in theory at least, provide safer home sites, a crowded community of celebrity homes on Malibu beach is as risky a place to live as any Gulf Coast barrier island jammed with beach cottages (see table 10). The indifferent sea doesn't discriminate between a rich person's mansion and a poor person's hovel.

Engineering Solutions to Rising Seas

"Hard" engineering solutions hold the shoreline in place with seawalls, groins, offshore breakwaters, or anything else that is static. These are appropriate for coastal cities (e.g., downtown Boston or lower Manhattan), but hard solutions will eventually destroy the beach for any beachfront community. Beach and dune replenishment, basically dumping sand imported from another location, is a "soft engineering" approach that can be quite expensive (typically a million dollars or so per mile [1.6 km]) and must be repeated every 3 to 9 years, depending on the location. Replenishment is neither a long-term nor a long-distance solution.

The Future of Some American Cities

- **Seattle.** This city faces a potential uptick in future flooding because of recent Puget Sound development on the narrow strip of land

Table 10. US Cities at Risk of Storm Surge Flooding by 2050

City	Number of People Affected
New York, New York	426,000
Hialeah, Florida	204,000
Miami, Florida	154,000
Fort Lauderdale, Florida	127,000
Pembroke Pines, Florida	120,000
Coral Springs, Florida	119,000
Miramar, Florida	100,000
St. Petersburg, Florida	91,000
Davie, Florida	90,000
Miami Beach, Florida	87,000
Charleston, South Carolina	83,000
Pompano Beach, Florida	80,000
Sunrise, Florida	79,000
Hollywood, Florida	76,000
Miami Gardens, Florida	72,000
Norfolk, Virginia	66,000
Lauderhill, Florida	66,000
Cape Coral, Florida	66,000
Boston, Massachusetts	62,000
Tamarac, Florida	60,000
Virginia Beach, Virginia	58,000
Tampa, Florida	57,000
Fontainebleau, Florida	56,000
Margate, Florida	53,000
Kendale Lakes, Florida	51,000

Note: These numbers tell us how many evacuees we might see by 2050 from rising seas and storm surge. Source: Adapted from Climate Central 2017.

between the shoreline and the coastal bluff, land made all the more hazardous by sea level rise. Bluff slumping during rains or winter storms is common.

- **San Francisco.** This metropolis has a problem because the city is growing in population while San Francisco Bay is expanding from sea level rise. The city assumes a 66-inch (1.7 m) sea level rise by 2100 and requires developers to take this into account when siting developments. If climate change continues its course (without human intervention), the bay will eventually expand east, turning Sacramento into a coastal or near-coastal city.

- **Houston.** Unfortunately, this major city has developed without zoning or planning. Houston is an outlaw city in terms of the extreme hazards it knowingly faces and ignores. This clearly is not acceptable in a time of major climate change. Houses have been built in areas known to flood frequently, even in emergency flood retention reservoirs, where over 5,000 homes were flooded by Hurricane Harvey in 2017. The city is in danger of a big storm pushing up the Houston ship channel and carrying highly polluted water. The higher the sea level, the greater the danger of a massive storm surge. A $33 billion sea gate and seawall (collectively known as the Ike Dike) planned in Galveston, Texas, is expected to prevent this. This could be the largest coastal engineering project in North America. Perhaps a sounder approach would be to spend a large part of the Ike Dike funds on cleaning up the ship channel.
- **Tampa.** In Tampa 31 local roads will flood in a storm surge or with a sea level rise of 18 inches (46 cm), blocking escape for those who respond too late to evacuation orders. A World Bank report noted (in a somewhat overblown statement) that Tampa Bay is "one of the 10 most-at-risk areas on the globe" (Tran 2013). "Lots of smoke but little fire" would characterize the community's efforts to plan for sea level rise.
- **New Orleans.** Sea level rise is not the most important hazard facing New Orleans. The biggest worry is another Hurricane Katrina destroying the levees and flooding the city. The US Army Corps of Engineers has done extensive levee repairs, which successfully kept Hurricane Ida (2021) at bay (so to speak). Decades down the road, the levees will have to be raised again. Local sea level rise is exacerbated by compaction and subsidence from groundwater extraction for domestic use.
- **Miami.** Among the world's 10 largest port cities, Miami is listed by the Organisation for Economic Co-operation and Development (OECD) as the city most endangered by flooding due to sea level rise (Nicholls et al. 2007). Despite the warnings, large upscale construction projects continue unabated in Miami. Of particular concern is the underlying permeable limestone, which is expected to allow flooding from below as the sea rises. The US Army Corps of Engineers is currently researching an estimated $8 billion, 13-foot-high (4m) seawall, in spite of the fact that seawater will flood the city

by moving up through the porous limestone, wall or no wall. A 2-foot (0.6 m) rise in sea level (likely by 2060?) will push groundwater into hundreds of thousands of residential septic tanks, rendering them unusable and contaminating Miami's bays and beaches with raw sewage. The Turkey Point Nuclear Power Plant will be cut off from the mainland (accessible only by boat). And as the sea continues rising, Turkey Point will be increasingly vulnerable to storm surge damage. And then there's the climate gentrification issue. Little Haiti, home to many of the city's low-income Black and Afro-Caribbean communities, sits on some of the highest land in South Florida. Residents there are coming under increasing pressure to sell their land to developers looking for the last bit of high ground to make a final profit before the inevitable collapse of the real estate market as the sea overruns South Florida.

- **Charleston.** According to NOAA, the sea level has risen 10 inches (25 cm) here since 1950. The downtown (the peninsula) already has a flooding problem, which will only increase with each jump in sea level rise. The city has installed pumps to remove floodwaters and has emplaced drainage pipes with check valves that let rainwater out but keep tidal and small-storm waters out. There is a federal proposal to build a $2 billion seawall 8 miles (13 km) long to protect the historic district. It is a certainty that Charleston will someday, in response to a higher sea level, become a walled city like some modern-day medieval castle town, surrounded by a moat made of ocean.
- **Norfolk.** In contrast to Houston, this city at the other extreme of climate planning has zoned itself to encourage development in safe areas and to discourage development in dangerous (frequently flooded) areas. One zone in the city that can't be protected from floods will still allow innovative development.
- **New York City.** New York has found its way thanks to Hurricane Sandy (2012). The plans for facing a rising sea include (1) constructing the Big U, a U-shaped berm around Lower Manhattan; (2) waterproofing Lower Manhattan by raising buildings or abandoning the first two or three floors; (3) placing generators on higher floors (beyond the sea's reach); (4) waterproofing utilities; (5) making a 40-mile (64 km)-long chain of barrier islands offshore (a ludicrous idea); and (6) building a 4,000-foot-long (1,220 m) living breakwater consisting of rock with a salt marsh and oysters planted behind it. Another proposal is to build a massive seawall between Sandy Hook, New Jersey, and Queens, New York, costing up to a whopping $119 billion.

Figure 45. Much of the downtown Boston waterfront, shown here, is at low elevation, sited on fill dredged from the adjacent bay. The city will need extensive protection from the rising sea in the near future. Photo courtesy of NPS / M. Woods.

- **Boston.** Much of Boston is threatened because it sits on extensive low-elevation fill made from mud and sand dredged from Boston Bay (see figure 45). A giant sea barrier costing tens of billions of dollars that would extend for 4 miles (6.4 km) between Hull and Deer Islands is the current proposed solution. The wall is intended to hold back high tides and storm surges. The plan seems unsuited for a high sea level rise of 8 feet (2.4 m).
- **Other places.** Other American coastal areas adversely affected by rising seas include Honolulu, Anchorage, North Carolina's Alligator Wildlife Refuge, the Everglades, the Mississippi River Delta, and the nation's 405 barrier islands.

What to Do

- **Location, location, location!** Don't live in a location at risk of being flooded by the sea. If you now live in a low-lying coastal area, you might want to relocate under the guiding mantra: Elevation, elevation, elevation!
- **Don't live on a barrier island.** Remember that a 3-foot (0.9 m) rise in sea level will end virtually all barrier island development because of frequent flooding in low-lying areas. Moreover, island residents will take on increasing financial burdens to pay for beach nourishment.

- **Don't live along estuaries** or bodies of water with tides.
- **Get a tide table** that lists king tides. Pay attention to potential storm surges coinciding with high tides, spring tides, or king tides.
- **Use the NOAA Sea Level Rise Viewer** (https://coast.noaa.gov/slr) to see how much your coastal community will be flooded by rising seas.
- **Elevate access roads.** What good is a high-elevation home site if your access roads are low and susceptible to storm surges or tidal flooding?
- **Live near evacuation routes.** See "Hurricanes" in "Air" for more about coastal evacuation.
- **Stabilize the shoreline.** Construct seawalls that hold the shoreline in place to attempt to protect buildings and people. This is an expensive approach better suited for a wealthy city like New York than for the countless smaller, middle-class communities edging the American shoreline. Seawalls generally destroy beaches, so there's an environmental cost as well.
- **Float cities.** This adaptation strategy employs floating structures that rise with the rising sea. (See "Green Cities" in "Space".)
- **Retreat.** Most coastal communities will eventually be forced to move away from the advancing ocean. This means physically moving homes and businesses to higher ground or abandoning them altogether and moving to new structures in a safer location as a family or as a business. Both NOAA and the Army Corps of Engineers agree that moving back must be part of any long-term coastal management solutions to sea level rise. A corollary to retreat would be to discourage development on or near beaches.

Ocean Acidification

If you're overfishing at the top of the food chain, and acidifying the ocean at the bottom, you're creating a squeeze that could conceivably collapse the whole system.—Carl Safina

The Evil Twin

Ocean acidification is often called climate change's evil twin. Both are born from an excess of carbon dioxide (CO_2) in the atmosphere. Both are rapidly growing threats to humanity and to Nature. Ocean acidification (a term first coined in 2003) is a relatively new problem. Its importance has only recently been recognized.

According to the National Oceanic and Atmospheric Administration (NOAA 2021), ocean acidification "puts the United States' $1-billion dollar shellfish industry and hundreds of thousands of jobs at risk." But the impact of acidification is global. Millions of people worldwide harvest seafood as a source of income, and billions rely on seafood as a primary source of protein. The United Nations (UN) 2021 climate report notes that ocean acidification could be a factor in global famines and the collapse of ocean-based economies (IPCC 2021a).

About 30% of the CO_2 that is supplied to the atmosphere by humans or that is released by melting of permafrost and from other sources is absorbed by the sea. This is around 22 million tons (19.95 metric tons) of CO_2 removed every single day from the atmosphere. This considerably limits the amount of global warming, but it comes at a heavy cost to the sea.

The Great Dying

Ocean acidification was part of the greatest mass extinction ever, at the end of Permian time, 252 million years ago (see "The Lessons of Geologic Time" in "Earth"). The Permian extinction (also called the *great dying*) extinguished 80% to 90% of all marine species. (Some sources put the total extinction of marine species at 96%). The problem was an infusion of massive amounts of CO_2 into the atmosphere, not from burning of fuel by industry and vehicles, but rather from enormous flows of lava. The lava responsible for the CO_2 emissions at the end of the Permian was the basalt flows east of the Ural Mountains known as the Siberian Traps (*trap* is a geologic term denoting a large accumulation of volcanic rock). At one time, the Siberian Traps may have covered an area the size of Australia!

But there was more than CO_2 responsible for the greatest die-off ever in the sea. Andrew Knoll in his book *A Brief History of Earth* (2021) referred to it as a deadly trio of a warming sea, lowering oxygen content of seawater, and ocean acidification, all occurring simultaneously. Methane emissions also played an important role in warming the planet.

A while back, scientists believed that moving CO_2 from the atmosphere and dissolving it in seawater was a good thing because it would reduce greenhouse gas warming. This is because tiny plankton incorporate CO_2 from seawater into their calcium carbonate shells. When the plankton die and sink to the seafloor, their shells are stored in the sediment. Thus, carbon is "permanently" removed from the atmosphere and the sea.

Over the past decade, however, it has become painfully apparent that the transfer of CO_2 into ocean water comes at a high price—the acidification of the sea. This is because CO_2 reacts chemically with seawater to form carbonic acid (H_2CO_3), begetting dire consequences for organisms made of calcium carbonate ($CaCO_3$) and for other marine life as well.

Scientists quantify acidity using the pH scale. The scale ranges from 0 to 14 with 7 being neutral. Numbers higher than 7 are considered alkaline (basic), and numbers below 7 are acidic. According to NOAA, the average pH in the ocean today is about 8.1, or slightly alkaline (NOAA 2020). But as the sea absorbs more CO_2 in the coming decades, the pH will decrease, and the oceans will become increasingly acidic and corrosive.

Since the Industrial Revolution, we have been casually pumping more and more carbon into the atmosphere. The ocean is a carbon sink that pulls carbon from the air, making ocean water more acidic. The pH of seawater has dropped from 8.2 to 8.1. This may seem like a minor change, but because the pH scale is logarithmic, that drop in pH represents a 30% increase in the acidity of seawater. The UN IPCC report (2021a) predicts that pH levels in the sea will decrease by 0.3 to 0.4 units by 2100. By the end of this century, the acidity of the world's oceans could therefore reach a catastrophic pH of 7.8. This would represent a 40% increase in acidity over preindustrial times. As one might expect, a similar process of acidification is also happening in freshwater lakes and rivers.

Impact on Marine Life

Calcifiers (animals that secrete calcium carbonate) like oysters, crabs, clams, corals, and mussels are at risk from ocean acidification for two reasons: (1) higher levels of carbonic acid in the ocean mean fewer carbonate ions are available to form shells, and (2) the increasing acidity of seawater will actually dissolve or partly dissolve those shells. So it's a double whammy for calcareous sea life. Carbonate is harder to get, and at the same time calcium carbonate shells are dissolving.

Apparently, oyster larvae in acidic water delay formation of their shells for around 48 hours after birth, during which time the acidic waters eat away at their tiny and fragile new shells. This is believed to be the cause of massive oyster die-offs in the Pacific Northwest (Waldbusser 2013).

Acidification particularly impacts organisms with thin calcareous shells like the free-swimming snails called *pteropods* (aka *sea butterflies*). These tiny organisms (about the size of a small pea) are an important food source for krill, salmon, mackerel, and even some whales (see figure 46). In 2011 NOAA researchers were shocked to find that more than half the pteropods found in shallow waters off America's West Coast showed signs of dissolution (Kennedy 2014).

Changing ocean chemistry also affects noncalcareous marine life. Scientists are just beginning to understand how a lowering pH apparently disrupts salmon's sense of smell, making it harder for them to detect predators and potentially to find mates. Salmon also need their sense of smell to locate their natal waters, the river where they were born. As the ocean pH continues to decrease, further disrupting salmon's olfactory sensitivity, some already declining salmon species may disappear from our rivers entirely.

Scientists expect ocean acidification will intensify with increasing concentrations of CO_2 and, when coupled with warmer waters and lower oxygen levels (which are also products of climate change), will eliminate most of the world's coral reefs by the latter part of this century. Warming makes

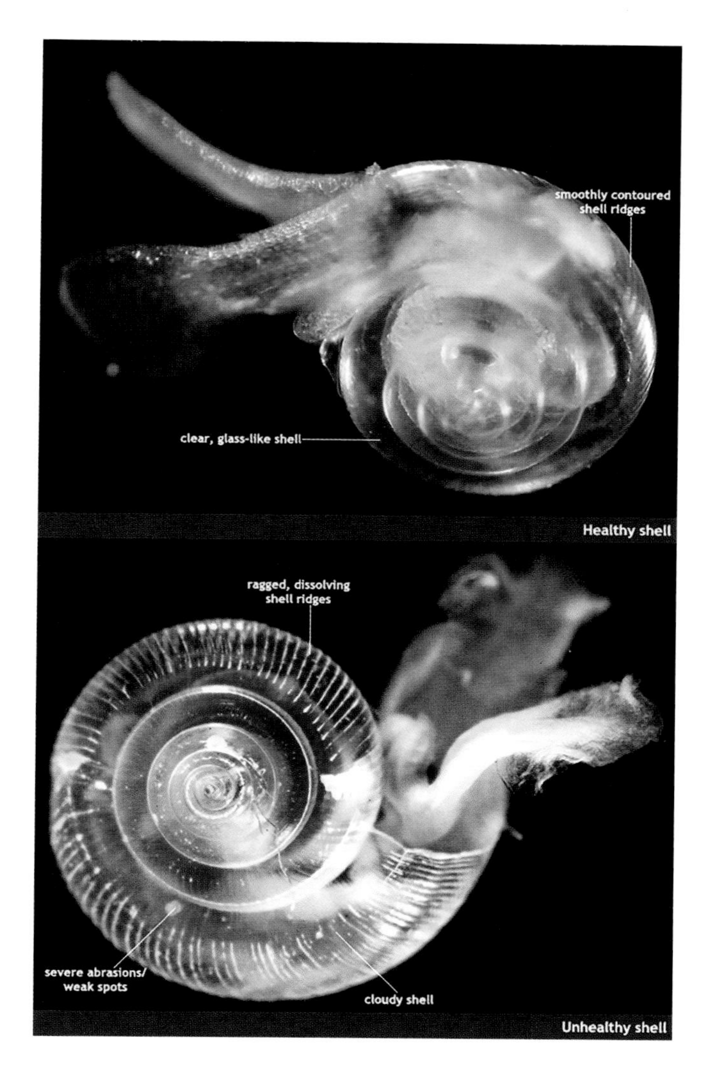

Figure 46. The healthy pteropod (*top*) lived in the laboratory for 6 days in pH-neutral seawater. The one on the bottom lived in acidic water for 6 days and formed the white lines of dissolution. Some are calling ocean acidification the "osteoporosis of the sea." Image from NOAA.

oxygen less soluble and thus will reduce the oxygen content of seawater to the detriment of life in the oceans. From 2009 to 2018 alone, the world lost an estimated 14% of its coral. The degradation and possible extinction of coral reefs, home to 25% of all marine life, has depressing implications not just for the health of marine ecosystems, but for human society as well. Think global food shortages and ruined tourism and fishing industries.

And the Good News Is . . .

There is some good news regarding our planet's oceans. While marine biodiversity in general is declining due to climate change, some sea snails are actually building thicker shells in response to increased acidity. The same appears to be true of the blue crab and the American lobster. How marine life responds to changing ocean chemistry varies from species to species and is largely unknown at present. Some will fade away and go extinct. Others will adapt and prosper.

Other good news . . . higher levels of CO_2 in the sea enhance the growth of marine algae, which like plants on land, can act as carbon sinks. This will not stop the acidification, but it may slow it down. In addition, scientists at the University of California, Davis, have demonstrated that cows emit 82% less methane when a small amount of seaweed (*Asparagopsis taxiformis*) is added to their diet (Nelson 2021). It seems that cows that eat seaweed burp and fart less and therefore emit less methane. Now scientists are exploring the feasibility of raising free-ranging cattle (and perhaps other ruminants) on a seaweed-enhanced diet.

What to Do

- **More research.** We need to do more research on the impact of acidification on specific species, from the standpoint of how acidification affects both ecosystems and fisheries.
- **More recognition: take acidification more seriously.** As a society, we need to recognize ocean acidification as another critical reason besides global warming to reduce greenhouse gas emissions.
- **Olivine to the rescue.** Francesc Montserrat and colleagues at the Netherlands Institute for Sea Research have suggested that spreading the mineral olivine along beaches will reduce the degree of acidification (Montserrat et al. 2017). Olivine in seawater exchanges magnesium ions for protons, thereby increasing the pH and reducing the acidity. Spreading this mineral globally is not possible, but perhaps it could be applied locally in the sea, in lakes and rivers, and theoretically on land.
- **New careers, new lifestyles.** As acidification takes its toll on the world's oceans, as reefs degrade and fish stocks decline or move elsewhere, communities sustained by wild-caught seafood will need to switch to farm-raised seafood or find other means of sustenance. Commercial fishermen and tour guides plying coral reefs will need to seek other kinds of employment. Similar adjustments may be required for those dependent on freshwater fisheries impacted by climate change and acidification.

Marine Heat Waves

Some of these islanders dutifully recited for us their ancient law: "Take no more from the sea in one day than there are people in your village. If you observe this rule, the bonito will run well again tomorrow."—Jacques Cousteau, *The Human, the Orchid, and the Octopus*

Death in the Pacific Northwest

It was a perfect storm of evil happenstance, the lowest low tide of the year coinciding with the most intense heat dome ever in the Pacific Northwest. Temperatures the day before reached 108°F (42.2°C) in Seattle and 116°F (46.6°C) in Portland, records for both cities. The June heat dome of 2021 would eventually lead to an estimated 1,400 heat-related deaths in the United States and Canada (Popovich and Choi-Schagrin 2021). But the toll on marine life was even more stunning in its magnitude. One billion marine organisms are thought to have died off Canada alone, with similar numbers for the coasts of Washington and Oregon, in large part because extremely low tides occurred around midday, exposing shellfish to intense sunlight. Barnacles, mussels, snails, clams, limpets, periwinkles, and other shellfish were cooked alive by the unrelenting heat. Scores of anemones, sea stars, crabs, and whole communities of intertidal creatures perished. In this fashion, the people living along the shores of the Pacific Northwest were introduced to a new extreme weather phenomenon—marine heat waves (Ma 2022).

Hot Water

The word Earth is something of a misnomer. Seventy-one percent of our planet's surface is covered by that magnificent and mysterious entity we call the ocean. Home to a million or more species (most of which are unknown to science), the ocean produces more than half the world's oxygen, regulates our climate by moving heat from the equator to the poles, and directly or indirectly provides food and income for hundreds of millions of people. It also absorbs as much as 30% of our carbon dioxide (CO_2) emissions while capturing an estimated 90% of the excess heat generated by those emissions. Why should we care about the health of the ocean? Because we live on "planet Ocean," and our lives and the lives of a million marine species literally depend on its health.

Marine heat waves are a newly recognized occurrence, or at least a newly named occurrence. They happen in all the world's oceans, including the Mediterranean. Underwater spikes of unusually warm temperatures can last for days or even years, wreaking havoc on fisheries and marine biodiversity. Between 2008 and 2019, the kelp forests off Northern California decreased in size by 95% because of hot ocean waters. It doesn't help that a wasting disease is killing off the sunflower sea stars that normally prey on sea urchins. Without predators to control their numbers, the urchins are spreading fast, devouring the kelp beds, their favorite food (Derham 2021).

There is growing global evidence that marine heat waves kill fish larvae, cause massive die-offs of seabirds, and allow the spread of unwanted invasive species. Marine heat waves are also degrading the world's coral reefs, bleaching entire reef systems as corals lose their life-sustaining symbiotic algae. The last global bleaching event started in 2014 and continued well into 2017, spreading across the Pacific, Indian, and Atlantic Oceans.

The primary cause of marine heat waves is ocean currents that build up large areas of hot water, already warmed by atmospheric heat. In times of light winds, surface waters don't mix with cool deeper waters so the warmed seawater remains in place until stronger winds arrive. It appears that marine heat waves are becoming more intense and more frequent (no surprise given that atmospheric CO_2 levels continue to rise). According to the Marine Heat Waves International Working Group (an international coalition of scientists formed to study marine heat waves), eight of the last ten most extreme events have occurred since 2010.

In order to raise public awareness, the working group proposed adopting a convention for categorizing marine heat waves, similar to the Saffir-Simpson Hurricane Wind Scale. Using their system, the Northwest Pacific Ocean event would have been a Category III marine heat wave, while the

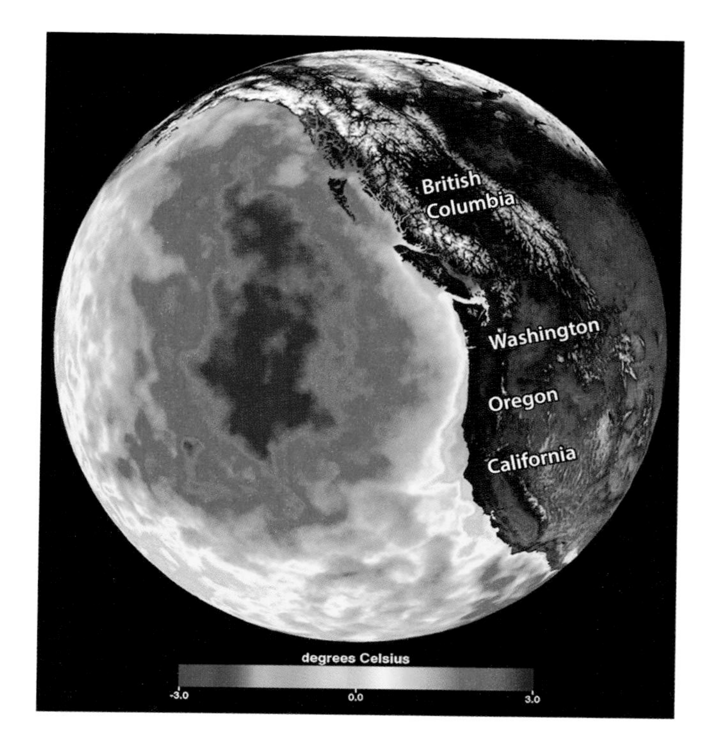

Figure 47. The 2015 marine heat wave known as the Blob covered the northeastern Pacific from Baja California to Alaska. The red tones represent the sea surface temperature anomaly (departure from normal for a particular location and time of year). Image from NASA Physical Oceanography Distributed Active Archive Center, modified by Charles Pilkey.

2011 Western Australia marine heat wave reached Category IV. The 2023 North Atlantic marine heat wave saw the highest ocean surface temperatures in the North Sea since the first records were first in 1850—an event NOAA listed as a Category IV marine heat wave.

The "Blob"

Between 2013 and 2016, scientists at the National Oceanic and Atmospheric Administration (NOAA) used satellite imagery, instrumented buoys, samples dredged from research ships, and hours of laboratory analysis to document a marine heat wave in the Pacific Ocean, where water temperatures were at times 5.4°F (3°C) above normal. By 2015 the heat wave covered 1,544,408 square miles (4 million km^2) of ocean, stretching from Baja California to Alaska's Aleutian Islands (see figure 47). Nicholas Bond, a climate scientist at the University of Washington in Seattle, dubbed it "the Blob," and this name stuck (Bond 2021). By 2017 it was confirmed that there had been an incredible 70% reduction in the cod harvest in the eastern Pacific in just a few years—an economic loss of $100 million annually (Cornwall 2019; NOAA Fisheries 2019).

The Blob started in 2013 with a high-atmospheric-pressure zone over the Pacific that blocked winter storms from mixing surface and deep waters. Over land this event (the Blob) would probably be called a heat dome. Beginning in 2014, sea temperatures were kept high by winds blowing warm water closer to the coasts of Oregon and Washington. The 2015 to 2016 El Niño made matters worse, broadening the extent of the Blob. The heat wave finally broke when La Niña brought back cooling storms and lower water temperatures.

Fish can swim to cooler depths, but some other marine species can't escape the hot ocean surface. During the Blob, the food chain was disrupted when krill (tiny shrimp that are a key food for fish and whales) disappeared from the coasts of Washington and Oregon. Toxic algae bloomed along much of the West Coast in 2015, killing more marine life. As food stocks declined, scientists believe the heat affected the metabolism of animals, forcing them to eat more to keep their bodies fueled. An estimated 1 million seabirds (common murres) starved to death, their bodies washing ashore by the thousands in Alaska, Washington, and Oregon (Birkhead 2020). Several species of whales also died. Scientists believe the loss of food and the increased metabolism ruined the Pacific cod industry.

Where Have All the Crabs Gone?

Marine heat waves are a sign of the sea's failing health in much the same way that a high fever is a sign of sickness in humans. The consequences of ignoring climate change are staggering, both for marine diversity and for our lives as well. In the fall of 2022, the king crab and snow crab seasons were, for the first time ever, canceled by the Alaska Department of Fish and Game (Bernton 2022; Xiang 2022). This was due to a precipitous decline in crab populations in the aftermath of a 2019 Bering Sea warming episode. More than 1 billion snow crabs mysteriously disappeared. Scientists speculate the crabs may have moved into deeper, cooler waters or perhaps were decimated by disease or the marine heat wave. No one knows, but climate change seems the most likely culprit. In any case, the sudden collapse of the $200 million crab industry will put immense economic stress on those who fish for a living (Bernton 2022; Xiang 2022).

Too often we forget that the health of the ocean and our own health (as well as our economic well-being) are fundamentally linked, are in fact one and the same. In a sense, we humans are both patient and doctor when it comes to an ailing ocean, victims of a self-inflicted sickness and custodians of its cure. Physician, heal thyself!

What to Do

- **Accept no easy fixes.** There is nothing to do, no adaptation strategy, no magic technological wand we can wave to make marine heat waves disappear. We must reduce global carbon emissions or else suffer the consequences. Some marine species are able to swim into deeper, cooler water. Some species like squid are evolving smaller bodies to survive warmer ocean temperatures. Shellfish in tidal pools will either adapt or die. The same is true for the seabirds.
- **Adapt to the changing oceans.** For some reason, California market squid (*Doryteuthis opalenscens*) have increased fivefold over the past two decades. Their population surges from California to Washington have been statistically linked to the occurrence of marine heat waves. As cod and other fish populations decline with warming waters, fishermen might try harvesting squid instead—but not off the coast of California, where the market squid is leaving in droves, following warming waters north toward Alaska (Turner 2022).
- **Keep the kelp.** Kelp beds have been compared to rainforests in terms of biodiversity and productivity. The kelp forests should be globally protected as marine Nature preserves. Divers are starting to cull sea urchins in the kelp beds off the US West Coast. Seals used to prey on the urchins, but the seal population has declined in recent years. Perhaps we can reintroduce seals from other places to keep the sea urchins in check.
- **Monitor and control harvests.** State and federal agencies should carefully monitor declining fish stocks and be willing to cancel harvests when needed to allow marine populations to recover. Those who harvest, transport, and sell the fruits of the sea should prepare for future cancellations, all but guaranteed to occur in the fast-warming seas of tomorrow.

Tsunamis

It was the biggest earthquake ever known to have struck Japan, and the fourth most powerful in the history of seismology. It knocked the Earth six and a half inches off its axis; it moved Japan thirteen feet closer to America. In the tsunami that followed more than 18,000 people were killed. At its peak the water was 120 feet high.
—Richard Lloyd Parry, *Ghosts of the Tsunami*

Geologic Surprises

In southeastern Alaska near the town of Petersburg, a barnacle-encrusted pine log sits on the side of a mountain at an elevation of 120 feet (36 m). Such an occurrence is a sure sign of a massive tsunami, probably caused by the failure of a nearby cliff. The cliffs there are still unstable and may someday collapse again, likely precipitating another tsunami. Scattered in the fjords of Alaska are hundreds of similar unstable slopes, and the chances for their collapse will only increase as the climate gets warmer.

Tsunamis are large waves usually initiated by earthquakes. But they can also be caused by volcanic eruptions, landslides, glacier calving, meteorites, and even weather events. Tsunamis generated by the collapse of a cliff or large landslides are relatively common, especially in the northern latitudes. The initial reason for the collapse is typically a local earthquake. But rocky slopes are losing structural support as glaciers at their base retreat and permafrost melts, a problem worsened by heavy rains that used to fall more often as snow. Tomorrow's landslide tsunamis will be generated by a com-

Figure 48. The outlined area on the mountainside is the portion of the slope believed to be unstable, in large part because the Barry Glacier has retreated up Barry Arm away from the base of the slope. When the slope fails, the resulting landslide will likely create a tsunami that will hit Whittier, Alaska. Photo © Gabriel Wolken, Alaska Division of Geological and Geophysical Surveys.

bination of tectonic movements creating earthquakes and slope instability due to climate change.

Tsunamis are among Nature's more impressive natural disasters. Recent notable landslide tsunamis in Alaska occurred at Taan Fjord in 1967 and at Grewingk Lake in 2015. The Taan Fjord tsunami reached a height of 633 feet (193 m). The one at Grewingk Lake was nearly 200 feet (61 m) high. Over the past century, similar landslide-induced mega-tsunamis (all 150 feet [46 m] or higher) have occurred in Greenland, Chile, Canada, and Norway (Higman et al. 2019).

Alaska's next mega-tsunami could possibly affect Whittier, a small fishing and tourist village along the shores of the steep-walled Barry Arm fjord in Prince William Sound. Because the Barry Glacier, which was at the foot of the slope, has retreated more than 2 miles (3.2 km), the unconsolidated slopes on the sides of the fjord have lost their protective ice anchor and are showing signs of incremental slippage (see figure 48).

Geologists expect the slope 30 miles (48 km) northeast of Whittier to fail catastrophically, possibly within the next 20 years. Such a collapse will trigger a tsunami that may overwhelm Whittier much like the tsunami from Alaska's 1964 Good Friday earthquake did, which killed 13 townspeople and did $10 million worth of damage. Today the potential landslide is being monitored by satellite and airborne sensors as well as by sensors attached to the ground—an alarm system that hopefully will never be used (Alaska Department of Natural Resources 2022).

The idea that global warming can trigger tsunamis is perhaps surprising to some. But our changing world may produce other geologic surprises as well. A growing body of evidence links climate change to increased frequency of minor earthquakes and perhaps of volcanic eruptions. Scientists in Taiwan have noted that small earthquakes occur when a typhoon passes over the island. They theorize that a typhoon's lower air pressure releases just enough weight pressing down on the crust that faults, already stressed to the point of breaking, suddenly give way. As typhoons increase in number and perhaps in power, earthquakes may do the same. Likewise, as glaciers in Iceland and Antarctica melt, volcanoes that were kept in check by the weight of overlying ice may start to erupt. Geologists are debating how serious these threats actually are. But the public should be aware that climate change's growing list of unanticipated events will likely include landslides, tsunamis, minor earthquakes, and increased eruptions from subglacial volcanoes as the ice melts (McGuire 2012).

Megatsunamis

The deadliest tsunami in recorded history (caused by clashing tectonic plates and unrelated to global climate change) occurred in 2004 and affected much of the Indian Ocean basin, killing perhaps 230,000 people in 14 countries. The maximum wave height is thought to have been around 100 feet (30 m), diminishing to 5 feet (1.5 m) when the waves arrived in South Africa (NOAA 2016).

The largest tsunami ever measured occurred in Lituya Bay, Alaska, in 1958. A massive landslide following a 7.8 magnitude earthquake generated a wave more than 1,700 feet (520 m) high, as indicated by a line of destroyed trees and bushes (called by geologists a *trim line*) on a mountainside. Two people in a fishing boat disappeared in the wave and were never seen again. Amazingly, a captain and his son in another fishing boat survived even though the tsunami lifted their boat hundreds of feet into the air. Close to the mouth of the bay, a couple in a third boat, lifted high by the crest of the wave, reported looking down at treetops far below. They escaped in a skiff after their boat foundered when it landed in the ocean. At least four other tsunamis are thought to have occurred in Lituya Bay since its discovery by the French explorer Jean-François de Galaup in 1786 (Miller 1960).

But no tsunami in human history comes close in sheer destructive power to the one off the Yucatán Peninsula 66 million years ago. That's when the Chicxulub asteroid (or comet), 6 miles (10 km) in diameter, crashed into the sea and formed an initial wave nearly a mile (1.6 km) high. Smaller tsunamis rippled across all the world's oceans. The Chicxulub Impact Event absolutely devastated the biosphere, eventually wiping out 75% of all life on Earth, including most of the dinosaurs (Brannen 2017).

Meteotsunamis

On July 4, 1929, an estimated 20-foot-high (6 m) wave crashed over a pier in Grand Haven, Michigan, suddenly and without warning. Ten people were pulled into Lake Michigan and drowned. Then, on June 26, 1954, while people were walking, fishing, and otherwise enjoying a sunny day on Chicago's lakefront, it happened again. An 8-to-10-foot-high (2-to-3-m) freak wave suddenly rose from a calm Lake Michigan and washed over Montrose Harbor, drowning eight people (Briscoe 2019). The origins of the waves in both tragedies completely baffled scientists. There were no earthquakes, no landslides, no significant winds, and certainly no volcanoes in the region. What could have caused such an event? Years later, after studying similar rogue waves around the world, scientists came to recognize a new kind of tsunami, one generated not by tectonics but by weather. They named the waves *meteotsunamis* (short for *meteorological tsunamis*). Figure 49 shows a 2018 me-

Figure 49. A Lake Michigan meteotsunami crashes over a breakwater on April 13, 2018, near the Ludington, Michigan, Breakwater Lighthouse. Photo by Debbie Maglothin, NOAA.

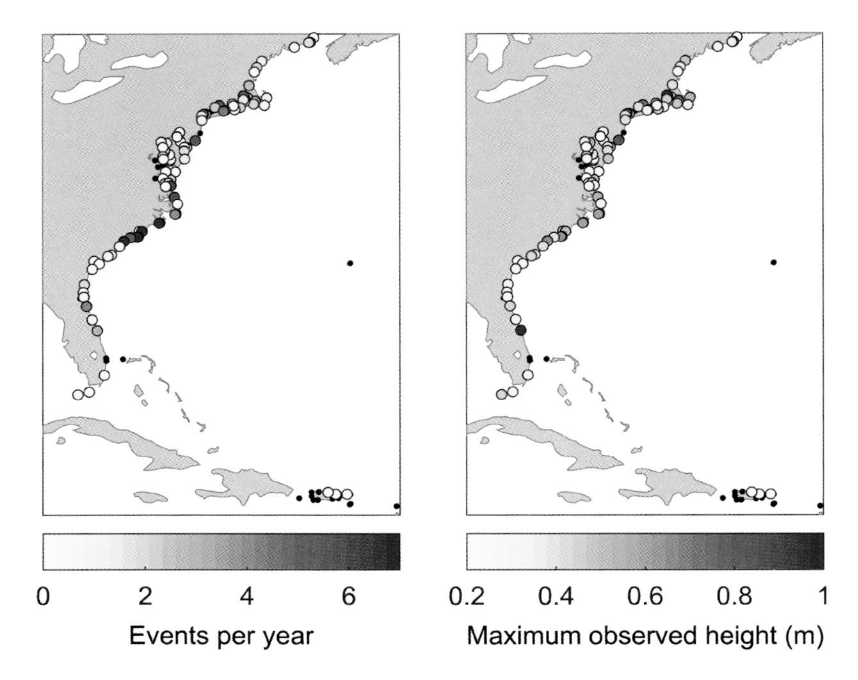

Figure 50. Locations, numbers, and wave heights of meteotsunamis on the US East Coast. Meteotsunamis are generally (though not always) small waves driven by intense weather events rather than by earthquakes or landslides. The black dots are where no meteotsunamis have been recorded. Image from NOAA, in Dusek et al. (2019).

teotsunami breaking outside the harbor at Ludington, Michigan. Historically large waves formerly attributed to earthquakes, storm surges, and so forth are now being reclassified as meteotsunamis.

Known locally by different names (e.g., *abiki* in Japan, *šćiga* in Croatia), meteotsunamis resemble conventional tsunamis in that they have similar wave periods but are caused by abrupt changes in air pressure accompanied by strong winds from squalls, derechos, thunderstorms, and the like. They are usually less than 2 feet (0.6 m) high, but when they coincide with shallow water, incoming tides, and other waves advancing into a narrow harbor, meteotsunamis can be amplified into large and potentially deadly waves.

Scientists as early as 1935 noted that weather can create tsunami-like waves. Only recently have they realized how pervasive and destructive such waves can be. The largest accurately recorded are the 19.5-foot (5.9 m) meteotsunamis that hit Vela Luka, Croatia, in June 1978, swamping boats and flooding houses, schools, hospitals, and factories. One year later, a 16-foot (5 m) meteotsunami rolled into Nagasaki harbor (Japan) and killed three people.

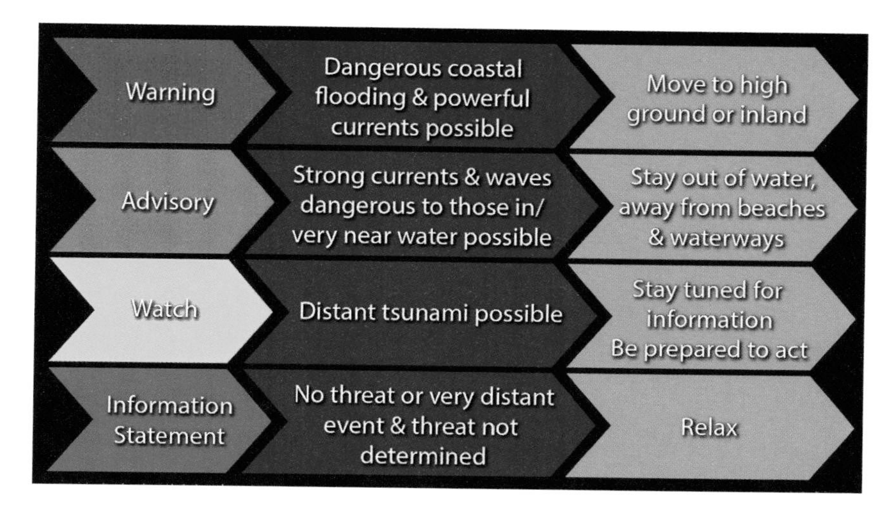

Figure 51. Tsunami alert levels. Image from National Weather Service.

Meteotsunamis are a global phenomenon and can happen on any seacoast or on the shorelines of large lakes. Unlike a seiche (pronounced "saysh"), a kind of standing wave that sloshes back and forth across a bay or lake, a meteotsunami moves as a series of waves (like a tsunami) in one direction (see figure 50). On average, 126 meteotsunamis occur every year in the Great Lakes and about 25 per year on America's East Coast (NOAA 2022). As the climate heats up, the frequency of meteotsunamis is expected to increase while sea level rise will make future meteotsunamis more deadly (Irwin 2021).

Scientists at the National Oceanic and Atmospheric Administration (NOAA) and in Europe are trying to find ways to forecast meteotsunamis and create a warning system similar to the tsunami alerts currently used in the United States and Japan. In the meantime those who live on the coast would do well to keep tabs on extreme weather events.

What to Do

- **Move** to higher ground away from the water if:

  You FEEL a strong earthquake.

  You SEE the water withdraw or white water advance an unusual distance.

  You HEAR a strange rumble or roar.

- **Listen** to tsunami alerts that can be heard on:

  NOAA's US Tsunami Warning System

  NOAA weather radio

  Marine VHF (channel 16)

  Radio, television, and internet

  Tsunami siren

- **Know** the tsunami alert levels (see figure 51).
- **Stay alert.** NOAA and other organizations are researching ways to predict meteotsunamis (NOAA 2021). For now we all need to be aware that meteotsunamis are a rare, but real, threat and that they can occur along any coast bordering a large body of water.

Floods

But water always goes where it wants to go, and nothing in the end can stand against it. Water is patient. Dripping water wears away a stone. Remember that, my child. Remember you are half water. If you can't go through an obstacle, go around it. Water does.—Margaret Atwood, *The Penelopiad*

The Power of Water

Any geologist worth her salt knows well the power of water and its temporal relations; the prodigious force of floodwaters that can uproot villages in a few minutes; the relentless energy of storm waves that reshape shorelines in mere days; the corrosive power of saltwater that over decades rusts cars, kills forests, and ruins a farmer's crops; the persistence of water over geologic time that flattens entire mountain ranges, washing their muddy remnants into the sea. Those who aspire to a happy life in peaceful coexistence with a watery biosphere had better learn to respect the power of water.

Climate Normals

Every 10 years, the National Oceanic and Atmospheric Administration (NOAA) recalculates an analysis of the past 30 years of weather data collected at weather stations across the United States to establish the "US

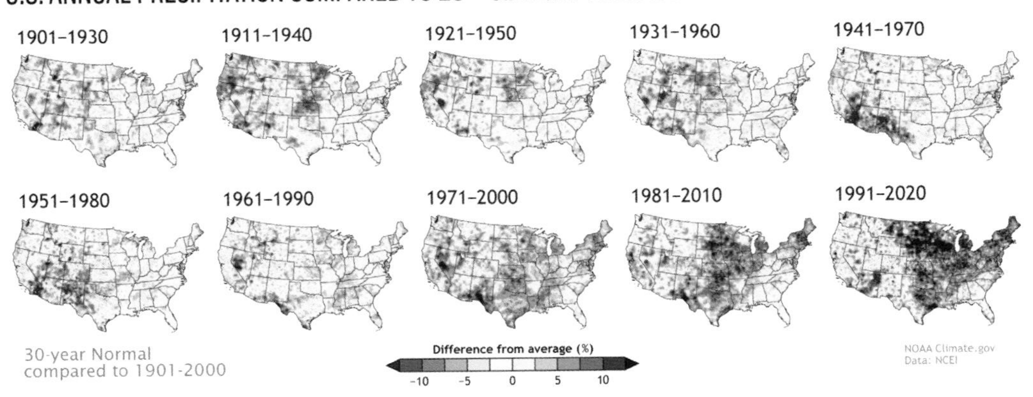

Figure 52. This figure paints a compelling portrait of how much wetter the United States has become as it has grown hotter. The last four maps in the figure, for 1961–1990, 1971–2000, 1981–2010, and 1991–2020, show a big increase in precipitation. Image from NOAA, climate.gov.

Climate Normals." As a cooperative member of the World Meteorological Organization, the United States is required to calculate "normals" at these intervals. This provides a baseline for comparing today's conditions to 30-year averages, telling scientists how much the climate is changing over time. These normals are then introduced to Americans through the daily weather reports on the news channels. In May 2021 NOAA released their latest data, showing that the United States today is generally hotter and wetter than in previous decades. The four most recent maps in the series (see figure 52) depict the four wettest periods. According to NOAA, at least some of that increased precipitation is linked to overall climate change.

Robert Sandford, a senior policy adviser for the Adaptation to Climate Change Team at Simon Fraser University in British Columbia, explained in 2021, "Hydrologic stationarity is the relative stability of the hydrologic system that we've relied on for the stability upon which we built our civilization," adding, "And we've lost hydrologic stationarity" (quoted in Smith 2021).

Sandford was interviewed in the wake of the extreme weather events that devastated British Columbia in 2021. The western Canadian province suffered the greatest combination of weather catastrophes that year in North America. The province was dried out, scorched, burned, and flooded, more or less in that order. On June 29, 2021, in Lytton, British Columbia, the temperature reached 121°F (49.6°C), the highest ever recorded in Canada (see

Figure 53. A bridge on the Nicola River along Highway 8 in British Columbia, one of many that failed or were damaged during 2021's catastrophic rainfalls. Photo © Province of British Columbia. All rights reserved. Reproduced with permission of the Province of British Columbia. British Columbia Ministry of Transportation and Infrastructure.

"Heat" in "Air"). The next day, 90% of the small town of Lytton burned to the ground in one of the 1,522 fires in the province that year. In total, wildfires burned more than 2,125,000 acres (860,000 ha) in British Columbia. A heat dome in June killed more than 800 residents and at least 651,000 farm animals (Labbé 2021). Next came intense rainfall from a series of seven atmospheric rivers in the fall of 2021, creating the "flood of floods." Floods and/or landslides breached most major highways and bridges (see figure 53). Wastewater plants were destroyed, and farm fields turned into swamps. We've entered a new era—not a new normal, but a new abnormal.

Sandford insists that what occurred in British Columbia is "far more staggering" than the usual flood damage we see on the nightly news. The truth is that we've located our farms and communities in places where the climate that used to be relatively stable is now starting to unravel. "What should be deeply alarming to all," Sandford said, "is that climate we have taken for granted for so long is, right before our very eyes, being replaced by a climate that, unless we act now, we may not survive" (quoted in Smith 2021).

With a warmer, wetter atmosphere comes an inevitable uptick in extreme precipitation events worldwide. In August 2021 severe thunderstorms struck Middle Tennessee, dropping an astounding 17 inches (43 cm) of rain in a

24-hour period, breaking the state record for daily rainfall. The National Weather Service had predicted moderate rainfall. But storms developed again and again, moving over the same area of Tennessee, releasing as much as 4 inches (10 cm) of rain per hour (Bacon and Yu 2021). The floods killed 20 people, 19 of them from the small town of Waverly. Footage from a home security camera captured just how quickly the town flooded. In just 12 minutes, it changed from pools of standing water on the road to a torrent of rushing water powerful enough to sweep away parked vehicles (ABC News 2021). It is noteworthy that this extreme event in Tennessee was not caused by a hurricane. Just a few days after the Tennessee flooding, Hurricane Ida, having made its way from Louisiana to New York, dropped a record 8.4 inches (21 cm) of rain, generating massive flooding in New York City, killing 13 people, including some who were trapped in basement apartments and drowned.

Recent intense rain events are certainly not limited to North America. As much as 8 inches (21 cm) of rain fell in less than 24 hours in Germany in July 2021. That same month, over 25 inches (63 cm) of rain fell in the city of Zhengzhou, China, in a 24-hour period, with nearly 8 inches (~20 cm) of the rain falling in just 1 hour. A year's worth of rain fell in just one day (Masters 2021).

Pakistan unfortunately has recently led a long parade of disastrous floods. Judging from satellite images, at least 10% of the entire country went underwater during the catastrophic monsoon rains in early June to late September 2022 (Miller et al. 2022). Thirty-three million people were directly affected by the floods, largely from the Indus River. Entire villages were destroyed, and more than a million homes lost along with some multistory buildings. All told, $10 billion of damage was done (Pratt 2022). The prime minister, Shebbaz Sharif, when pleading for aid from other countries, noted that Pakistan produced less than 1% of the world's carbon emissions (Ramirez and Dewan 2022).

Prime Minister Sharif is justified in reminding us that countries like Pakistan contribute little to climate change's causes yet suffer the most from its consequences. Should China, the United States, and the other carbon superpowers compensate countries for the blatant injustice of climate change? Should we reimburse African countries hit with years of drought followed by massive floods in 2023? What about Pacific Islanders who are losing their homelands to rising oceans? How much money is enough to compensate for national extinction? These are tough ethical questions. Some argue that, at the very least, we have a moral duty to countries like Pakistan to speedily reduce our emissions. Perhaps we should focus on solving the problem . . . not the apportionment of blame.

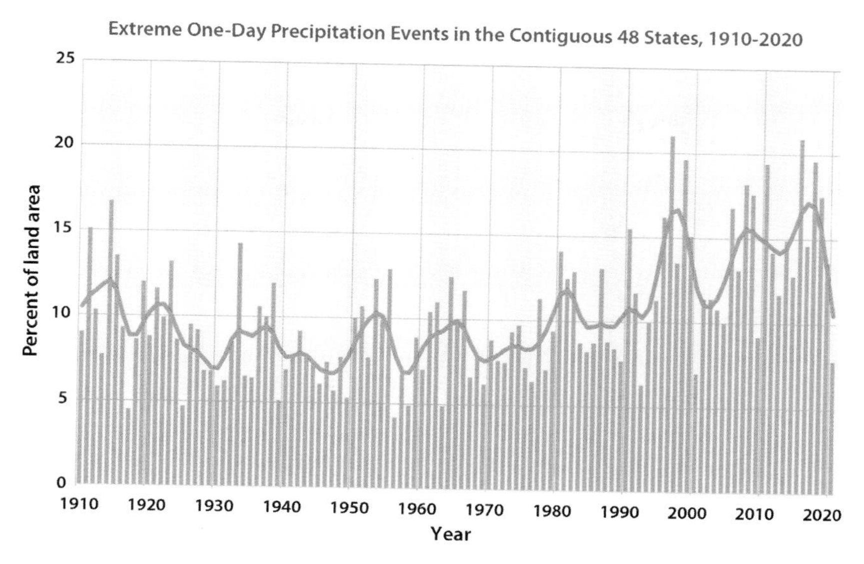

Figure 54. This figure shows the percentage of the land area of the 48 contiguous US states, where a much-greater-than-normal portion of total annual precipitation between 1910 and 2020 came from extreme single-day precipitation events. The bars represent individual years, while the orange line tracks a 9-year weighted average. Since 1996 the frequency of extreme events has increased, a reflection of climate change. Image from the EPA, NOAA.

Why So Wet?

A hotter atmosphere from climate change allows air to hold more water vapor, thus increasing the potential for extreme precipitation. A 1.8°F (1°C) increase in average surface temperatures means the atmosphere can hold 7% more moisture. The changing behavior of the jet stream also plays a role in triggering severe weather. If the jet stream is moving in a straight line, storms move along with it, but if the jet stream meanders, weather systems can stall, leading to extreme and potentially deadly weather (heat domes, heavy snowfalls, cloudbursts, floods).

Per the US Environmental Protection Agency (EPA), in recent years a higher percentage of precipitation in the United States comes in the form of extreme single-day events (see figure 54). The prevalence of these events remained fairly uniform between 1910 and 1980. Nine of the top 10 years for single-day extreme weather events have occurred since 1996.

Rivers in the Sky

Quick, what's the longest river in the world? The Amazon? The Nile? The Yangtze? Nope, trick question. The US Geological Survey (USGS) notes that the longest freshwater "rivers" in the world are atmospheric rivers (2021). Chances are, if you live on the US West Coast, you know all too well the term *atmospheric rivers,* as these phenomena have recently caused extensive flooding in the region.

Typically a half-mile (0.8 km) above ground and up to 5,000 miles (8,000 km) long and 300 miles (483 km) wide, atmospheric rivers are narrow (relative to their length) bands of water vapor that flow like a "river" in the sky (Weill 2021). When such a river passes over mountains the river is cooled. The water vapor condenses and precipitates as rain or snow. A large atmospheric river, like the ones that flow across the Pacific Ocean from Japan to California, can move water at a rate 15 times the average daily discharge of the Mississippi River (USGS 2021).

Atmospheric rivers span the western reaches of North America from southeastern Alaska to Baja California. They also occur in Western Europe, the west coast of North Africa, Iran, New Zealand, and both the east and west coasts of Australia. Those affecting North America that originate near Hawaii are known colloquially as the "Pineapple Express," a catchy sobriquet popularized by newscasters that belies their potential for deadly flooding.

Atmospheric rivers are infamous for the strong winds and widespread flooding they bring, but they can also bestow a measure of benevolence by alleviating droughts, restocking depleted reservoirs, and helping to extinguish wildfires. Unfortunately, the heavy rains they dump on recently burned, vegetation-free slopes can trigger deadly mudslides (see figure 55).

A 2022 study published in *Science Advances* predicts that over the next century, the US West Coast, British Columbia, and Colorado will see an increase in wildfires, followed by extreme rainfall delivered by atmospheric rivers, another one-two power punch delivered by climate change (Touma et al. 2022). The increased rains will challenge federal, state, and local governments. At the federal level, we can expect massive costs to the National Flood Insurance Program. State and local governments will have to invest heavily in repairing and improving infrastructure. A 2021 report from the First Street Foundation identified 2.2 million miles (3.5 million km) of road and 25% of critical infrastructure facilities as being at risk of flooding. Those numbers will increase over time.

Between December 20, 2022, and January 15, 2023, a parade of nine particularly powerful atmospheric rivers washed across the state of California, dumping up to 25 inches (63 cm) of rain in some places. The resulting floods, mudslides, and wind damage from "bomb cyclones" (atmospheric rivers that develop strong winds from fast-dropping air pressure) destroyed homes and killed at least 20 people. Some of these atmospheric rivers spawned storm systems that continued across the country, triggering floods and tornadoes

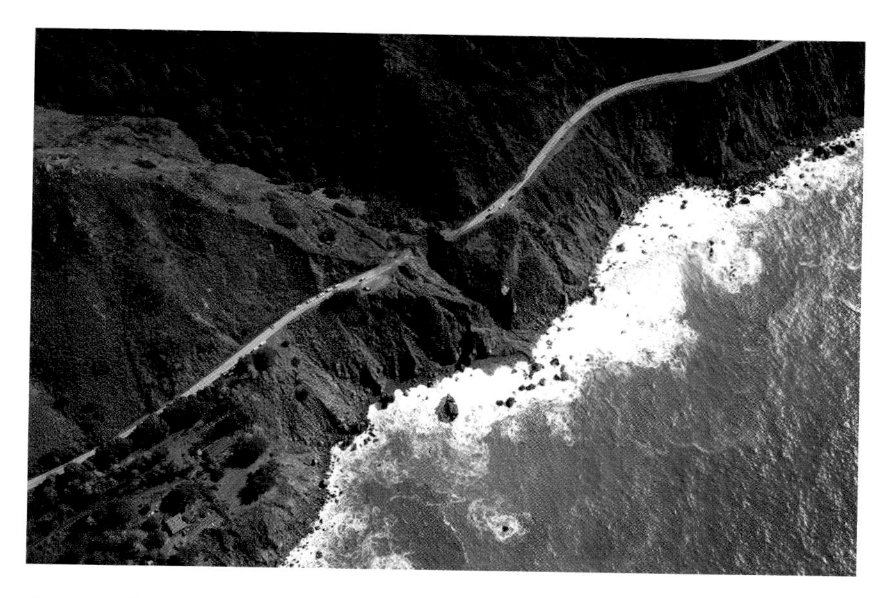

Figure 55. The January 26–28, 2021, atmospheric river along the central California coast hit these slopes that had been de-vegetated by the 2020 Dolan wildfire. With nothing to stabilize the slope, heavy rains washed dead trees, rocks, ash, and mud downslope, causing massive erosion and gully formation and damaging US Highway 1, California's coastal highway famed for its scenic beauty. Photo from the US Geological Survey.

in the Mississippi River Valley. The good news is that the deluge of rain and the 20 feet (6 m) of mountain snow dumped in the higher elevations of California did much to alleviate the state's drought, at least for now (Cappucci 2023). There was no good news for those living near the Mississippi River.

Scientists have developed a five-category scale (similar to the Saffir-Simpson wind scale for hurricanes) to quantify the intensity of atmospheric rivers. Category 1 means an atmospheric river is mostly beneficial, while Category 5 means it is mostly hazardous. As ocean surface waters get warmer, atmospheric rivers are expected to increase in intensity, leading to more Category 4 and Category 5 atmospheric rivers.

Californians worry about the next "big one," referring to the next major earthquake that will hammer San Francisco or Los Angeles. But the next big one may not be a seismic event at all but, rather, a catastrophic atmospheric river superstorm that will flood large portions of the state, causing hundreds of billions of dollars in damage and dumping the kind of rainfall seen once in a thousand years. At least, this is what some USGS natural hazard scientists think could happen should ocean temperatures continue rising. Those who live in the path of tomorrow's sky river superstorms should be wary. Expect nothing . . . and be prepared for anything, including the worst that a changing climate might throw at you.

Types of Floods

The NOAA National Severe Storms Laboratory identifies several different types of floods (NOAA, n.d.). River floods can come from excessive rainfall from tropical cyclones, monsoons, thunderstorms, snowmelt, ice jams, or dam or levee failure. Coastal floods involve inundation of coastal areas from higher-than-typical high tides, sometimes enhanced by wind, rainfall, and/or low atmospheric pressure. Storm surges are floodwaters generated by ocean storms like hurricanes. Storm surges can be particularly destructive when they coincide with high tides.

Sometimes inland floods occur with several consecutive days of moderate rains, or when intense rain falls over a short period, without the effects of a river overflowing its banks. Houses are inundated as rain piles up and seeps through the floorboards, a not-uncommon event in the Pacific Northwest.

And, finally, flash floods are caused by heavy rainfall over a short period. Flash floods can rip through riverbeds, canyons, or city streets (see figure 56). Areas that have been burned are particularly susceptible to flash flooding. Pay attention to weather alerts, and avoid areas predisposed to flash flooding.

Surprises on the Floodplain

The best way to avoid floods is to avoid living in flood-affected places (especially floodplains). Remember that extreme precipitation events might expand flooding beyond historic flood zones. If you reside in a flood-prone area and flood insurance is available, you should acquire it. But choosing a home at a higher elevation far from a floodable area makes good sense. Keep in mind that extreme precipitation may cause flooding in small creeks. The Germans were surprised when smaller creeks caused extensive and deadly flooding in 2021 after unexpected heavy rains.

The citizens of South Africa's KwaZulu-Natal province in April 2022 were equally surprised when catastrophic rain and floods swept away bridges, highways, and houses and killed over 300 people. In eastern South Africa, shipping containers from Durban's port were seen floating down roads, and an oil tanker washed up on Durban's famous tourist beach (see figure 57) along with a large python (Burke 2022). A dozen crocodiles escaped a crocodile farm during the storm. It could have been worse. In an earlier flood years before, 15,000 crocs escaped in the floodwaters.

Figure 56. The north entrance road, one of many destroyed in Yellowstone National Park by the heavy rains and massive snowpack melt of mid-June 2022, forcing evacuation of almost 10,000 tourists and temporarily closing this very popular park. Photo by Kyle Stone, US National Park Service.

Figure 57. In April 2022, catastrophic rainfall in South Africa's KwaZulu-Natal province killed more than 300 people, scattered hundreds of shipping containers (shown here), and grounded an oil tanker on Durban's tourist beach. A large python also arrived on the beach, apparently transported there by floodwaters. At least one shipping container was seen floating down a local road. As usual in such storms, bridges, highways, and buildings were destroyed. Photo from the EPA.

Glacial Outbursts

Climate change is happening so fast that every day a new, previously unrecognized climate event gets reported, or so it seems. According to a 2023 *Scientific American* article, 15 million people worldwide are at risk of being washed away in a flood caused by the rapid melting of mountain glaciers (Harvey 2023). Lakes made of meltwater form at the base of shrinking glaciers. Any rapid, precipitous melting of the ice can cause those lakes to overflow, pouring a torrent of water downslope, destroying houses, hydroelectric stations, farms, or anything else in the way. These "glacial lake outburst floods," as they are called, have already killed hundreds. India, Pakistan, Peru, and China are particularly vulnerable, but scientists fear the same kind of flooding will happen in the Alps and in the mountain glaciers of North America.

The mother of all glacial-lake outbursts occurred at the end of the last ice age when an ice dam formed across the Clark Fork River near what is now Missoula, Montana. The dam held back the waters of Lake Missoula, a glacial lake 2,000 feet (~610 m) deep that held a volume of water equivalent to Lake Erie and Lake Ontario combined. When the ice dam finally burst, the ensuing flood covered much of eastern Washington state and the Willamette Valley in Oregon. So great was the surge of water that ripple marks (still visible today) 50 feet (15 m) high were formed on the valley floor. Boulders as big as cars were floated in chunks of ice and deposited 500 miles (800 km) away. Eventually, the floodwaters joined the Columbia River. From there they roared to the sea, carving the sides of mountains along the way to create the Columbia River Gorge.

There were in fact people living in the Americas when the waters of Lake Missoula roiled across Cascadia. But any encampments along the Columbia River would have been swept away, and anyone who witnessed that outburst from close quarters would not have lived to tell the tale. Glacial outbursts have happened before. They will happen again.

Just when you thought it was safe to go near the water . . .

Avoid Floodwaters

Do not drive through floodwaters. According to the Federal Emergency Management Agency (FEMA), a foot (30 cm) of water can sweep even large vehicles off a road, and water depths are not always easy to determine. In addition, if you drive on a flooded road, you may not be able to accurately determine where the sides of the road end. If you find yourself in floodwater and the water is rapidly rising in your car, exit immediately, seek refuge on the roof, and call for help. If your car stalls and the floodwaters are not moving rapidly, abandon your car and seek higher ground. Remember the maxim TURN AROUND. DON'T DROWN!

Avoid walking or swimming in floodwaters. As little as 6 inches (15 cm) of moving water can knock you off your feet (FEMA n.d.), and murky waters may obscure dangers. Downed or underground power lines pose risks of electrical shock. In addition, floodwaters often contain pollutants, including sewage. You should never go into floodwaters if you can avoid it. In recent years, social media has featured videos of idiots diving from bridges into floodwaters and floating through the alleys of flooded cities. They may as well be swimming in a septic tank.

Environment Florida reported that in 2017 Hurricane Irma released over 28 million gallons (106 million liters) of wastewater across Florida, equivalent to every resident in the city of Miami flushing the toilet 38 times (Environment Florida, Florida PIRG, and Frontier Group, n.d.). A report from the Center for American Progress noted that the floodwaters from Hurricane Sandy spilled close to 11 billion gallons (42 billion liters) of raw or partially treated sewage into the streets, rivers, and coastal waters of New Jersey and New York, enough to fill the Empire State Building 14 times (Bovarnick, Polefka, and Bhattacharyya 2014). Floodwaters are no playground and should be avoided if possible.

Floods can be terrifying events. The 2022 flooding in Pakistan noted above started with a severe heat wave. Large volumes of meltwater from the country's 7,000 glaciers triggered minor flooding in the spring. Then the monsoons arrived early and lasted longer, dumping five times the normal rainfall. The rains fell continuously for more than 2 months. The Indus River swelled, overflowed its banks, and became a lake 62 miles (100 km) wide (Miller et al. 2022). Roads, bridges, farmers' crops, and entire villages were washed away. By November more than 1,700 people had died, including hundreds of children. More deaths are expected to follow as cholera, dengue, malaria, and other water- and vector-borne diseases take their toll (Pratt 2022).

Damage from climate-change-induced floods (like those in Pakistan) includes loss of topsoil and ruined crops, as well as damage to infrastructure. In an example of out-of-the-box thinking, New York City is distributing inflatable temporary dams to residents (especially those living in basement apartments) who may experience floods when heavy rains arrive. The dams can be inflated with water and are typically used to block floodwater from pouring down driveways. This plan relies on residents to show resilience and take personal action, action that will hopefully save lives (Weather Channel 2022).

What to Do

- **Calculate risk.** One tool available for prospective property buyers or owners in the United States is Risk Factor, a free online tool from the First Street Foundation (2022) that helps identify a property's risk of flooding. The tool provides FEMA flood zone information but also takes into consideration a location's risk of flooding from high-intensity rainfall, overflowing rivers and streams, and coastal storm surge. At riskfactor.com you can enter your address into a search engine, and within seconds it will categorize the overall risk of floods and the risk of flooding over the next 30 years. The website also identifies a property's risk of heat, wildfires, and wind.
- **Check before you buy.** Homebuyers should also check FEMA's online Flood Map Service Center. In many states, there is no legal requirement for a seller to report flood risks or flood history.
- **Announce glacial outbursts.** Governments need to alert people living near mountain glaciers of the newly recognized danger of glacial lake outburst flooding.
- **Rain watch.** The record of catastrophic rainfalls indicates that their magnitude usually comes as a surprise. Rain can be predicted but not a major rainfall. Best assume the worst when planning how to respond.
- **Be app savvy.** Use a weather app on your phone to stay aware of storm watches and warnings.
- **Monitor flash flood potential.** Be wary of potential flash flooding when hiking in the canyons of the West. Sometimes flash floods arrive without warning from distant thunderstorms, out of sight and beyond the range of human hearing.
- **Use disaster resources.** Ready.gov gives a broad range of advice about responding to any disaster, before and after.
- **Critter caution.** Be careful returning to a flooded home as animals, including snakes or even alligators, may have taken shelter inside.
- **Avoid mold.** If you have asthma or other respiratory problems, or are immunocompromised, avoid exposure to mold.
- **Make an emergency plan now.**

  1. Plan escape routes that don't go through flood-prone areas.
  2. You may want to purchase a kayak, canoe, or inflatable boat for escape during extreme floods that may cover roads.
  3. Assume gas stations and stores will be closed, so have an escape bag (see "Bug-Out Bags").

4. Do not walk, swim, or drive through floodwaters.
5. Consider purchasing a vehicle escape tool to keep in your glove compartment should you need to break a window to escape a flooded vehicle.
6. If possible, avoid bridges over very fast-moving waters.
7. Be sure your emergency plan includes your pets.
8. Move valuables in your home to higher levels.
9. Keep your rain gutters and drains clean and functional.
10. Don't forget your medications. It's also a good idea to take a list of your medications with you or have one accessible online.

Drought

They always say to Californians that we don't have seasons. Of course, that is not true. We have fire, flood, mud, and drought.—Phyllis Diller

Climate Change and Drought

Drought . . . the word conjures images of parched vegetation, mud cracks where lakes once stood, desiccated carcasses of livestock . . . a rainless land somewhere in East Africa perhaps, or maybe South Asia, but certainly not in the fortune-favored lands of the wealthy West. But climate change has made such perceptions outdated. Drought can happen anywhere, anytime. Indeed, it is already unfolding in the American Southwest and the northern half of Mexico, now suffering through their worst drought in more than 1,000 years (see figure 58).

Drought is a prolonged period of abnormally low rainfall leading to land degradation and a shortage of water. The key word in this definition is *abnormally*. Regions such as the Sahara Desert that are normally very dry with low rainfall are not necessarily suffering from drought, unless the little rain that does fall is abnormally low.

Unlike most states, Washington State has a legal definition of drought: "An area is in a drought condition when the water supply for the area is be-

Figure 58. Saguaro National Park, Arizona, near the Mexican border. The giant saguaro cactus (*Carnegiea gigantea*), native to the Sonoran Desert, is threatened by higher temperatures, drought, and invasive plants like buffelgrass (*Cenchrus ciliaris*) that increase the frequency of wildfires. Photo © Norma Longo.

low 75 percent of normal. Water uses and users in the area will likely incur undue hardships because of the water shortage" (Washington State Government 2023). Perhaps other states will follow the Evergreen State's lead as droughts increase in frequency and ferocity.

Average precipitation in the United States has increased 0.2 inch (0.5 cm) per decade since 1901. Global precipitation has increased 0.1 inch (0.25 cm) per decade. The problem with water security is not the amount of precipitation but climate change's impact on its distribution and frequency. There are other impacts as well (see figure 59).

Climate change triggers drought by (1) raising air temperatures, which increases evaporation from the soil; (2) diminishing snowpack and melting it sooner; and (3) changing Earth's weather patterns, pushing weather systems like atmospheric rivers north, for example, away from places that might need the rain. Scientists think that climate change will likely produce a future where wet places are wetter and dry places get drier. But there is a degree of uncertainty in this claim, as the computer models cannot accurately forecast future precipitation patterns. Modeling temperature as it correlates to changing CO_2 and water vapor levels (water vapor is also a greenhouse gas) by contrast, is reasonably accurate.

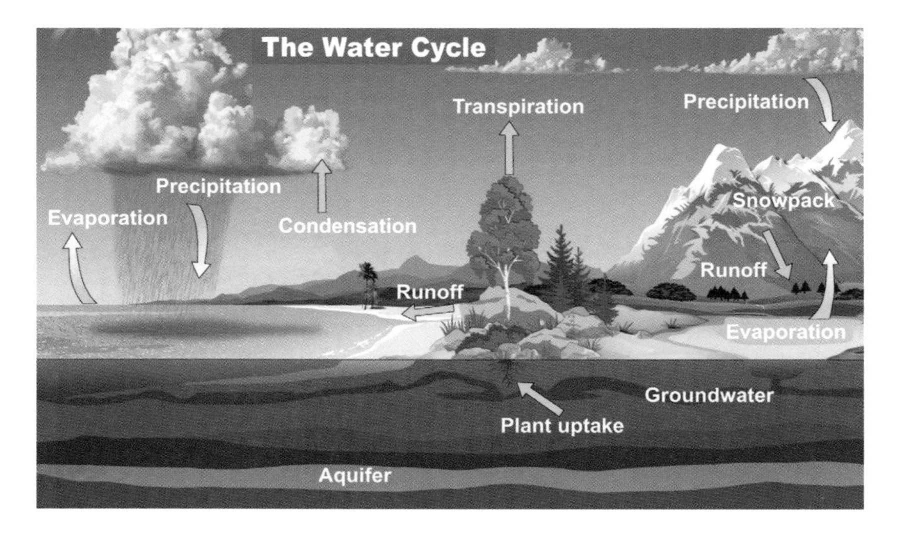

Figure 59. Water gets recycled on the surface of the Earth by evaporation, condensation, and precipitation. Climate change is modifying everything shown in the illustration. Image from NOAA, modified by Charles Pilkey.

Figure 59 illustrates the mechanics of the water cycle. In effect, a drought happens when something interrupts one or more of the components of the water cycle (i.e., precipitation, runoff, or evaporation).

Even a cursory study of drought reveals some of the many connections between climate change, ecosystems, and human society—what is sometimes referred to as the *web of life* but might better be simply termed the *biosphere*. Cogent examples abound: Global warming is making droughts more frequent and more expansive, which in turn dries soil and vegetation and increases the likelihood of wildfires, dust storms, topsoil loss, and desertification. Warmer winters are decreasing snowpack in both the Northern and Southern Hemispheres. The snowpack, whether thick or thin, reflects heat back into the atmosphere (known as the *albedo effect*). Less snowpack allows more sunlight to heat the Earth's surface, thereby exacerbating drought.

Moreover, the same drought that reduces household, farming, and industrial water supplies also reduces water levels in rivers and lakes, stressing wildlife and killing fish. Lower river levels concentrate pollutants, foster mosquito-borne diseases, halt river transport, and limit the amount of water available to suppress fires. This last item translates to bigger wildfires that could potentially destroy more of the world's forests and grasslands.

There are economic consequences to drought as well. Parched fields under empty skies can result in crop failures, higher food prices, financial distress for farmers, and economic stress for low-income families. Droughts

Table 11. Types of Drought

Meteorological drought	Prolonged period of low rainfall leading to a shortage of water.
Agricultural drought	Based on impacts to farming from deficits in rainfall and soil water, reduced groundwater, or reservoir levels needed for irrigation.
Hydrological drought	Based on impacts of rainfall deficits on water supply to stream and lake levels, reservoirs, groundwater, and power plant cooling.
Socioeconomic drought	When the demand for goods, e.g., foods, exceeds supply as a result of a weather-related deficit in water supply.
Flash drought	An unusually rapid-onset drought characterized by hot soil temperatures and low soil moisture. A flash drought can suddenly appear in weeks or even days.

Source: NOAA, National Weather Service.

worsened the Great Depression (ditto for the next big depression). If a drought persists long enough, reservoir levels can drop so low that hydroelectric plants have to be shut down (see "Water Supply"). Likewise, there may not be enough water in rivers and lakes to cool power plants (including nuclear plants). If the power goes . . . so goes the local economy.

Prolonged drought can trigger crop failures, food shortages, and in some cases widespread famine and death (see "Famine" in "Earth") not only for human communities but for wildlife as well. To the credit of our species, when presented with televised images of a drought-induced famine, the peoples of the world pool their resources to alleviate the suffering of the hungry . . . the hungry humans, that is. There are no food-aid concerts for hungry elephants.

Drought is a dark thread woven into the fabric of climate change. To better understand that thread, the US National Weather Service found it useful to arrange droughts into five categories based on a particular drought's characteristics or its impact on society. These are described in table 11 (see also figure 60).

Droughts Around the World

Climate change is increasing the likelihood of drought in large areas of the United States (see figure 61) and around the world. In the US Southwest (Southern California, Nevada, Arizona, Utah, and New Mexico), droughts are expected to become more intense, more frequent, and longer lasting. For the most part, the same is true globally.

Some of the more drought-prone countries in the world are Morocco, Uganda, Syria, Somalia, Iran, Pakistan, China, Afghanistan, Eritrea, Sudan, and Ethiopia (Kiprop 2018). Other countries going through drought in-

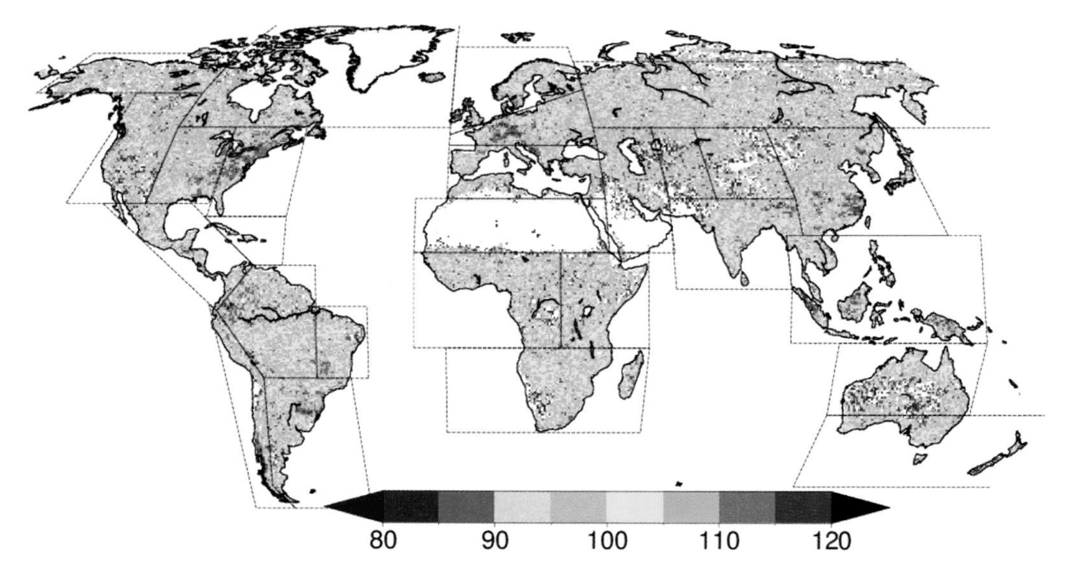

Figure 60. World map showing where flash droughts might occur. A flash drought is one that develops rapidly, in a matter of a few days. Any location in yellow, orange, or red is susceptible to flash drought. Image © Sourav Mukherjee and Ashok Mishra, *Geophysical Research Letters* 2022, https://phys.org/news/2022-02-hotspots-drought.html.

clude Portugal, Madagascar, and Chile. In Portugal, the non–life-threatening drought is hydrological, a winter phenomenon limiting water use from some reservoirs used for hydroelectric power and irrigation. In Madagascar, undergoing its worst drought in 40 years, people are dying from extreme famine. Chile's historic 13-year megadrought is the result of changing weather patterns. Warmer than usual waters off Chile's coast are blocking the storms that used to bring rain. Lake Peñuelas, the reservoir that once supplied Valparaiso with all its drinking water, has completely dried up and turned to desert.

Drought is a major reason that climate refugees from Syria, North Africa, and sub-Saharan Africa are streaming into Europe. Refugees from Central America seeking asylum in the United States are in many cases fleeing drought in their respective countries. The drought in Syria (2006–11) is considered to be a contributing cause of civil war as farmers, displaced from their land, crowded into the cities, where employment was nonexistent. No rain, failed crops, jobless farmers, political unrest, and armed conflict ultimately forced migration to greener shores. We've seen this movie before. We'll see it again.

Ethiopia has been suffering a particularly bad drought over the past 3 years. According to the United Nations Children's Fund (UNICEF), 6.8 million people, including 220,000 already malnourished children, need humanitarian aid because of dried-up water wells, failed crops, and dying livestock.

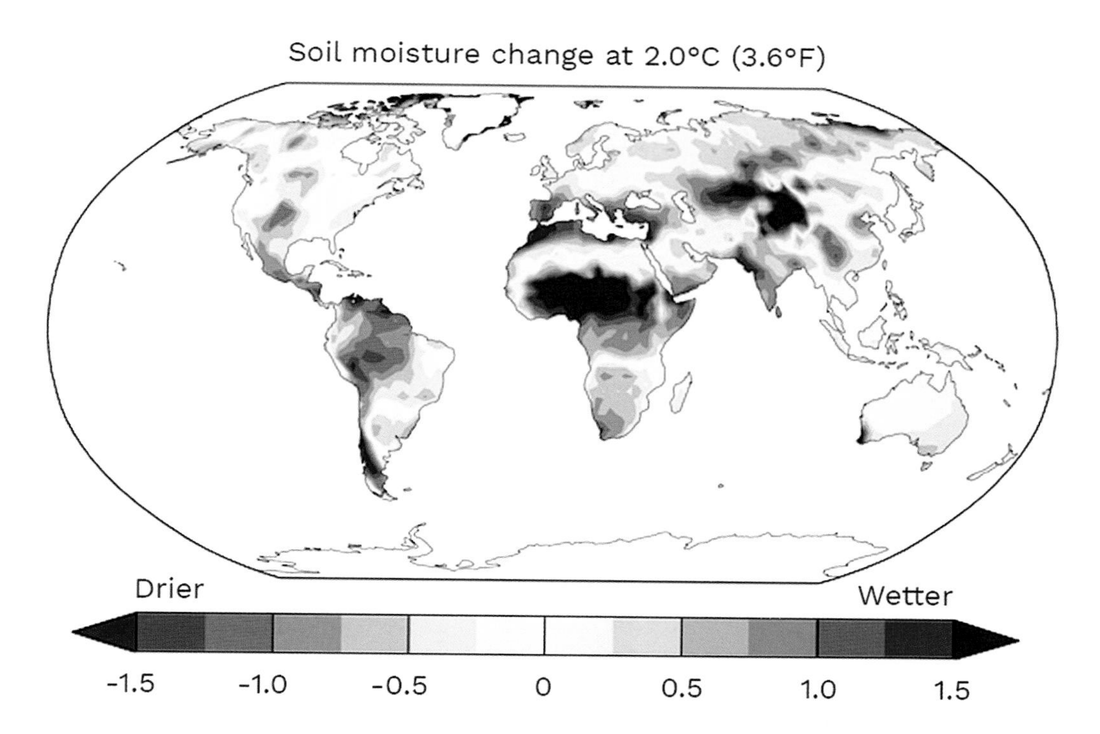

Figure 61. Projected global changes in soil moisture content with a 3.6°F (2°C) rise in temperatures. A 3.6°F rise in global temperatures could happen as soon as 2060. The brown shading represents areas with potential drought. Image from IPCC (2021b, 17).

The problems extend into Somalia. In Syria, Ethiopia, and recently in Sudan, the suffering from a prolonged drought is magnified by ongoing political conflicts (UNICEF 2022).

In 2022 China, Europe, and the United States were simultaneously hit with heat waves and severe drought. Sections of China's Yangtze River dropped to their lowest levels since 1865, curtailing hydropower and forcing the country to ramp up its coal-fired plants. As Europe endured its worst drought in 500 years, water levels on the Rhine slumped to the point where cargo ships had to carry lighter loads to avoid running aground. On the Mississippi, barges were lined up, waiting for the US Army Corps of Engineers to dredge channels deep enough for them to pass safely. These days the Big Muddy, it would seem, is getting a whole lot muddier.

The Dead Pool

In 2022 the ongoing drought in the US Southwest dropped the water level in Lake Mead (one of 15 reservoirs on the Colorado River) to its lowest point since the 1930s. If the lake goes down another 100 feet (30 m) or so, it will reach what is called *dead pool*, the point at which there is insufficient water to produce electricity. More than a million people get electricity from Hoover Dam. More than 25 million get water from its lake (NPS 2020).

Lake Mead's dead pool has recently acquired a different and altogether more sinister association. Human remains half-buried in the muddy lake bottom were exposed as water levels continued to recede. Several skeletons have been discovered so far. Victims of drowning, boating accidents, or foul play: the dark secrets of Lake Mead are slowly coming to light.

Forty million people in seven states depend on Colorado River water. Southern California alone gets about one-third of its water for drinking and farming from the Colorado River. Discussions for water distribution plans are ongoing, and some restrictions in allowable water volumes have been instituted. More will likely come in the future. In the spring of 2023 meltwater from abnormally high snowpacks in the Colorado Rockies alleviated (at least temporarily) the low-water levels on the Colorado River. Nevertheless, the long-range forecast for the Colorado River and for the American Southwest in general is for continued drought.

The future of Hoover Dam, Lake Mead, and the Colorado River is problematic. So much water is taken out of the Colorado River for irrigation, drinking, and other uses that this once proud river dries up in the Mexican desert, somewhere south of Yuma, Arizona, long before it reaches the sea. Declining water levels behind the Hoover Dam point to an uncertain future for the still-growing city of Las Vegas. What will the city do if the lake runs dry? (Hill 2022).

Desertification

One important impact of drought related to atmospheric warming is desertification. This means the formation of new deserts or the expansion of existing deserts (now happening in Central Asia and Africa). A desert is any large area of land with sparse vegetation and little precipitation. A third of the Earth's landmass is desert. Deforestation, improper land management, and overgrazing can also contribute to the expansion of deserts. See table 12 for a comparison of the sizes of the world's major deserts.

The Sahara Desert is the largest hot desert at around 3,500,000 square miles (9,065,000 km^2) in extent. There are cold deserts as well. Antarctica and the Arctic are the largest deserts in the world, but these two cold deserts are shrinking as the ice melts. Desertification is very likely enlarging many hot deserts; the Sahara is currently expanding south into Sudan and Chad (see figure 62). The deserts of western China are spreading on average

Table 12. Area of Major Deserts

North American Deserts	Area (square miles)
Great Basin Desert	200,000
Mojave Desert	48,000
Sonoran Desert	119,000
Chihuahuan Desert	175,000
For comparison	
Antarctic Desert	5,480,000
Arctic Desert	5,360,000
Sahara Desert	3,500,000

about 1,300 square miles (3,367 km^2) per year. To attempt to curb desertification, the Chinese government has sent soldiers into the areas of expansion to plant trees (Baker and Roubab 2019).

Feedback Loops

Global warming leads to higher temperatures which can lead to drought. Drought kills trees. When a tree dies, it releases CO_2 into the atmosphere. This results in more atmospheric warming, more drought, more dead trees and so on . . . a classic example of a positive feedback loop in which an initial change to a system leads to other changes that then enhance that initial change. An example of a negative feedback loop might be as follows: more CO_2 pumped into the atmosphere from the burning of fossil fuels stimulates plant growth. Increased plant growth sequesters CO_2 from the atmosphere. Less atmospheric CO_2 reduces global warming.

Unfortunately, of the 40 or so major climate change feedback loops that scientists have identified, nearly all are positive, meaning they will perpetuate and enhance climate change. It's possible there is a negative climate feedback loop out there in Nature that no one has thought of yet. But this seems unlikely, given what we know from the geologic past and from what is happening now.

There is, however, one possibility that may yet save the day, a climate feedback loop that scientists rarely if ever mention in their reports. As global warming intensifies more books like this one will be written. There will be more climate documentaries, news reports, and films. More citizen protests, letters to politicians, green technologies . . . all leading to a global effort to tackle climate change. We, the collective peoples of the world, are a de facto

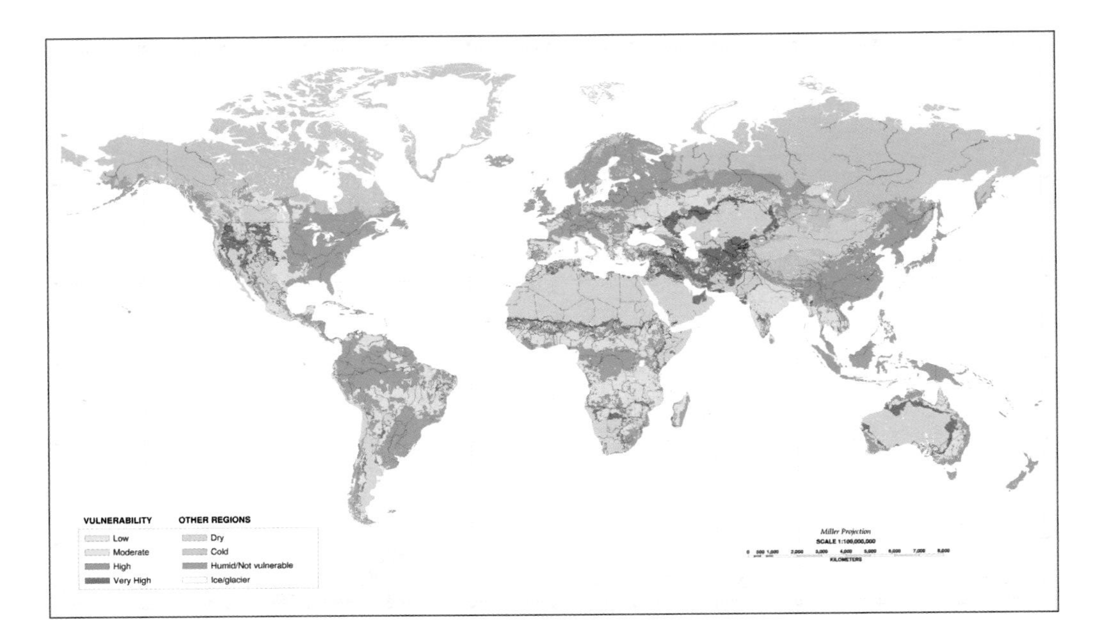

Figure 62. World map showing vulnerability to desertification. Gray areas are already largely desert. Red shading indicates areas that are particularly susceptible. Sub-Saharan Africa, Kazakhstan, Afghanistan, Iran, and parts of China, the western United States, and northern Mexico are regions that are currently undergoing desertification. Image from the US Department of Agriculture.

feedback loop, the best there is, and the only one we know that can return the world to some semblance of climatic stability. May it be so. . . .

What to Do

WATER CONSERVATION AT HOME

- **When it's yellow, let it mellow.** Avoid flushing the toilet unnecessarily. Dispose of tissues, insects, and similar waste in the trash, not the toilet.
- **Take short showers**, and don't take baths. Wash clothes, dishes, and your body less frequently and with less water.
- **Conserve water in the bathroom.** Avoid letting the water run while brushing your teeth, washing your face, or shaving.
- **Place a bucket** in the shower to catch excess water for watering plants.
- **Fix leaks** (immediately).
- **Install rain barrels** for emergency use and for watering plants. Seal with a fine mesh screen to keep mosquitoes from laying eggs.

- **Downsize your lawn.** Minimize watering your lawn. Or, better yet, shift to lawnless landscaping. Think Zen gardens.
- **Rocks in the toilet.** Put a rock in the toilet tank to reduce water usage.
- **Never throw away water**—find a use!
- **Bottled water.** Stock up on a 2-week supply of bottled water. Store water for pets. For long-term droughts, buy an emergency water tank.
- **Pets.** Drought can be dangerous for pets. Make sure your pets have plenty of drinking water.
- **Organize or participate** in community meetings about water conservation.

WATER CONSERVATION BY LOCAL GOVERNMENTS

- **Initiate water rationing** before drought gets severe.
- **Eliminate leaking pipes.**
- **Build underground storage tanks** to store water in anticipation of future droughts.
- **Rooftop water storage.** Require new buildings and retrofit old buildings to incorporate water storage areas on roofs to harvest rainwater.
- **Early warning systems.** Monitor weather, soil conditions, depletion of aquifers, snowpack, and reservoirs to help forecast drought (especially flash drought).
- **City climate plans.** Revise city and state climate plans to meet new climate change issues.
- **Made in the shade.** Plant native trees and shrubs, especially those that are drought resistant. Increased shade cools the soil and decreases soil evaporation. Shade also cools those who work outside, thereby reducing their water consumption. Sweat less and drink less but stay hydrated.
- **Sustainable farming.** Use sustainable farming practices to conserve water for irrigation. Examples include drip irrigation, drought-resistant crops, or soil covered with branches, twigs, weeds, etc., to reduce evaporation.

EMERGENCY PREPAREDNESS AT HOME

- **Solar still.** Water shortages from drought are happening all over the world. In addition to the aforementioned examples in Chile, Syria, and elsewhere, there are towns in South Africa that are counting the days until their water runs out, and families can access tap water for

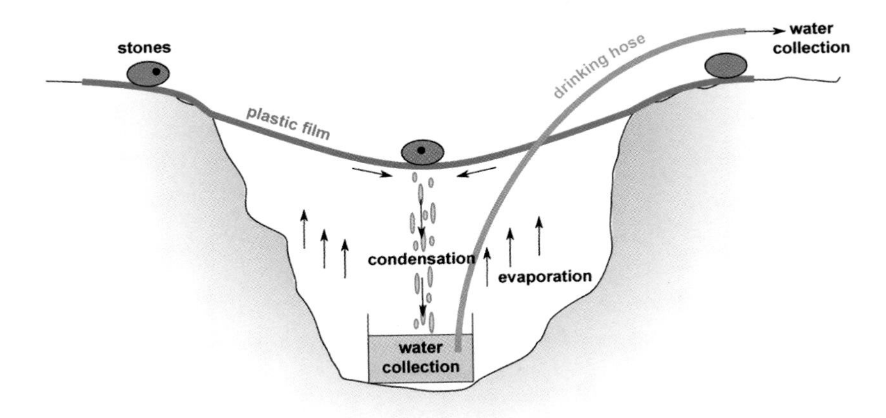

Figure 63. Solar stills distill water using the heat of the sun to evaporate impure water. The water is then cooled and collected minus its impurities. A primitive solar still, like the simple design shown here, might keep a family going with drinking water during a climate crisis. Even saltwater can be purified by this method. Commercial solar stills are available (from Amazon, among other vendors). Image by Daniele Pugliesi (GNU Free Documentation License, Version 1.2 or any later version published by the Free Software Foundation. This file is licensed under the Creative Commons Attribution-ShareAlike 3.0 Unported, 2.5 Generic, 2.0 Generic and 1.0 Generic license).

only a few hours a day in drought-ridden communities in Mexico. Whether from drought, power outages, war, or some other climate-related crisis, there may come a day when your community runs out of water. A primitive solar still, like the simple design in figure 63, might keep a family supplied with drinking water through the magic of condensation. By adding river water or seawater to the bottom of the pit, you can increase the amount of water collected in the pan. You can also buy commercial solar stills online.

Water Supply

Often, they sat in the evening together by the bank on the log, said nothing and both listened to the water, which was no water to them, but the voice of life, the voice of what exists, of what is eternally taking shape.
—Hermann Hesse, *Siddhartha*

The Water of Life

Water is magic. Liquid poetry. The stuff of rich metaphor . . . the stuff of life. For 10,000 years civilizations grew accustomed to a stable climate providing, beyond the occasional drought, a reliable supply of clean water. Climate change is about to upend that stability. The rains will still come. But they will fall in murderous abundance in some places and in other places not at all. Some farmers will benefit from increased rainfall averages. Others will curse empty skies and despair over disappearing snowpacks that once greened their fields with meltwater.

Through a complex interaction of anthropogenic causes, some related to climate change and some not, the quality, quantity, and geography of water have become a disquieting concern. Population growth and climate instability will lead to water shortages, diminishing water quality, collapsing ecosystems, and ultimately conflict, both within nations and across national borders. The water wars are about to commence.

Figure 64. The chart illustrates how water is used in the United States. Aquaculture means fish farms. Thermoelectric power uses coal, gas, oil, uranium, and so on to make electricity from steam-driven turbine generators. Domestic use accounts for 1% of the water used in the United States yet is the first item rationed by most state and city governments. Drawing by Charles Pilkey.

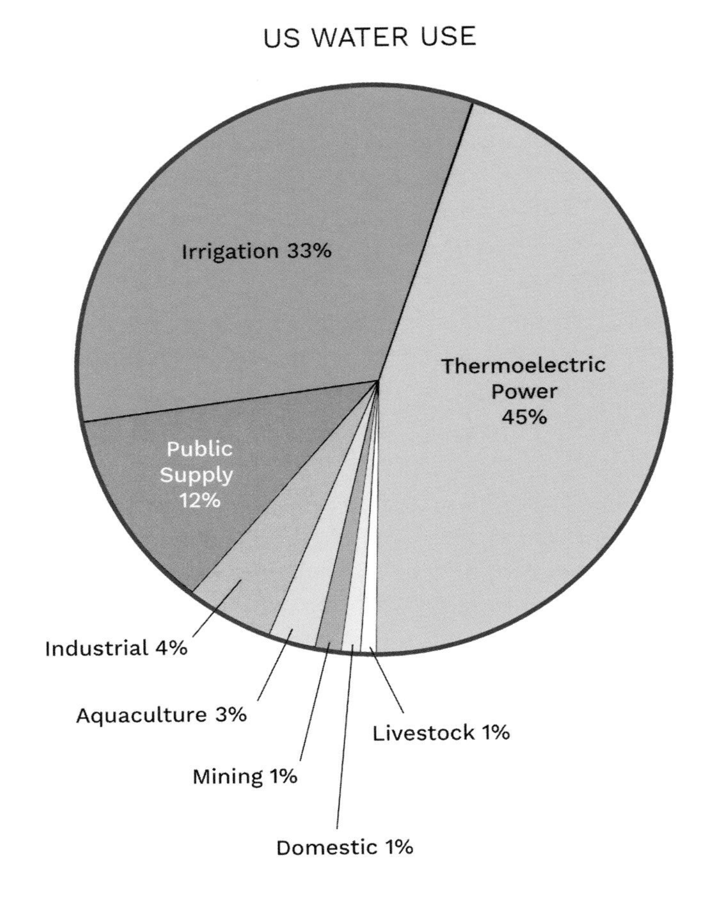

Water Use in America

The average American uses 156 gallons (590.5 liters) of water every day, including water for toilet and shower use. By contrast, the average person in India consumes 38 gallons (144 liters), while in Mali daily use per person is a mere 3 gallons (11.5 liters) of water (CDC 2020). Figure 64 shows water use in the United States. The percentages listed are a national average that don't reflect regional and local variations. A small farming community in Iowa, for example, will use more water for irrigation than a fishing village in Alaska.

Human population growth, deteriorating water quality, and climate change (see "Drought") will stress water supplies for cities and ecosystems in the coming decades. Figure 65 shows where competition and conflict are expected to occur between American cities and states by 2060.

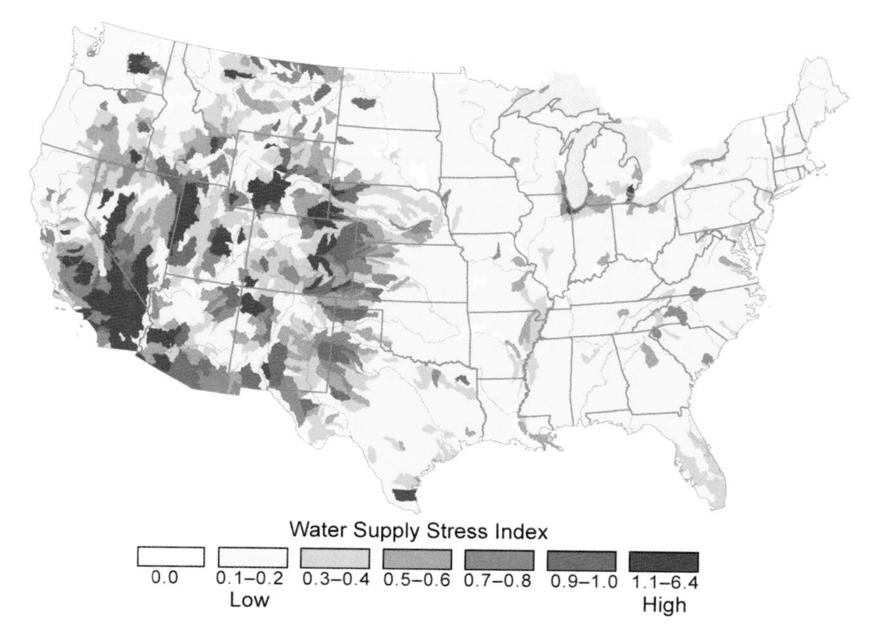

Figure 65. The map shows water stress in the United States in the early part of the 21st century, when demand exceeded 40% of available water supply. The brown areas are where conflicts over water use are likely to occur in the coming decades. Image from Averyt et al. (2011), 26.

Dead Zone

Eutrophication is a fancy term for what happens when rain carries an excessive load of nutrients from soil into rivers, lakes, estuaries, and coastal waters. The nutrients (mostly nitrogen and phosphorus) can come from natural sources like minerals in rocks but more often are from fertilizers for farms and lawns, partly treated sewage, and industrial wastewater. The nutrients cause algal blooms. The algae sink to the bottom, where they are decomposed by bacteria, a process that depletes water of its oxygen (hypoxia). Carbon dioxide (CO_2) and methane are by-products of the decomposition of algae. Fish, snails, shrimp, worms, and any oxygen breather that can't swim away in time suffocate. It's a problem at the mouths of major rivers and in other bodies of water like the Great Lakes, where eutrophication has created biological deserts known as *dead zones.*

The world's largest dead zone is in the Black Sea. The second-largest resides in the Gulf of Mexico off the coast of Louisiana. The biggest Gulf dead zone ever measured was 8,776 square miles (22,730 km^2) in 2017, roughly the size of Massachusetts (see figure 66). The area of hypoxia expands in summer and mostly dissipates after hurricanes churn Gulf waters, replenishing the sea with fresh oxygen. The cool waters of winter further reduce algal

Figure 66. Water is considered hypoxic—a dead zone—when its oxygen level drops below 2 mg of oxygen per liter of water. At that level most life either dies or swims away. The map shows the global extent of hypoxic ocean water. Virtually every coastline near a big city is hypoxic. Lack of oxygen combined with warming and acidification grows dead zones. Image from Breitburg et al. (2018). Map created from data provided by R. Diaz, updated by members of the GO_2NE network, and downloaded from the World Ocean Atlas 2009.

growth until spring when warming waters lead to eutrophication, reactivating the hypoxia cycle (NOAA 2021).

According to the Nature Conservancy, the Gulf dead zone costs the fishing and tourism industry an estimated $82 million a year (2022). Nationwide that number is $2.2 billion (Chislock et al. 2013). Commercial fishermen, already stressed by hurricanes and the BP oil spill, have to pay higher fuel costs to catch fish that elude hypoxic waters by swimming further offshore.

Sometimes algal blooms produce toxins that get into the food web, killing animals, hurting people, and even poisoning a city's water supply. This is what happened in Toledo, Ohio, in 2014, when toxins from cyanobacteria spiked in Lake Erie, Toledo's primary source of drinking water. For three days, 500,000 Toledoans were unable to use tap water for drinking, showering, and brushing their teeth or swim in the lake until the level of toxins finally shrank to an acceptable level. Runoff from leaking septic systems, cattle feedlots, and farms caused the toxic algal bloom, a perennial problem in the Great Lakes and in other lakes across the country (EPA 2018).

Ocean oxygen levels have declined globally by 2% since 1960, in large part because of climate change (Oschlies et al. 2018). In lakes the situation is much worse. Oxygen levels there have decreased 5.5% globally since 1980 (NSF 2021). Warmer water temperatures are partly to blame since warmer water holds less oxygen than cold water. But rising temperatures and higher CO_2 levels also stimulate algae growth, worsening oxygen depletion already happening from eutrophication. Nutrient runoff from increased flooding (especially when coupled with heat waves) means more eutrophication, less oxygen, expanding dead zones, and bigger fish kills. Declining water quality is another climate issue that will challenge optimism.

Snowpack

Snow that accumulates over winter in mountains is called *snowpack*. At higher elevations or near the poles, snowpack may never melt. Instead, the accumulated snow gets compressed over time to become a glacier. Melting snowpack replenishes creeks, rivers, and reservoirs and also provides water for farms and industry.

Because of global warming, the snow line is moving upslope, and snowpack is melting earlier around the world, sometimes melting so rapidly as to cause flooding. Vanishing snowpack is a problem in California, where one-third of the water used for drinking and irrigation comes from melted snow (Dimick 2015). It's even more critical in Asia, where at least 1.6 billion people depend on meltwater from the Himalayan snowpack and from glaciers (Dixit 2019).

In April 2015 California's snowpack was at 5% of its normal capacity (estimated to be the lowest in 500 years). For California's farmers, who produce a third of America's vegetables and two-thirds of the country's fruits and nuts, that year was a disaster (Spiegel 2021). During the growing season, when thirsty crops most needed water, there was none to be found. Meltwater from snowpack was already long gone. Creeks dried up, stranding fish in isolated pools. Over 21,000 jobs were lost, and agricultural revenue losses approached $3 billion. It was a good year to grow weeds (Stevens 2015).

The Alchemist

There are still rivers in the world where you can cup your hands and safely drink the water. But humanity's ever-expanding imprint will make such places harder to find. Dead zones in oceans and lakes; dried lakes turned to desert; microplastics in the sea, in tap water, and even in Antarctic snow . . . we seem to have lost the momentum gained by environmental policies initiated at the first Paris Summit in 1972 and with the passage of the Clean Water Act in the United States. Earth's hydrosphere is changing before our eyes, transmuted into some baser alchemy by a fast-shifting, shifty climate.

Most of us imagine sea level rise as a mass of seawater moving leisurely through centuries of sluggish time over the surface of the land. But this is a sleight of hand, a deceptive legerdemain by the trickster alchemist we call climate change. The sea is also rising from below, seeping into groundwater, contaminating wells, and poisoning farmers' fields with salt. Evidence of saltwater intrusion (also called *salinization*) can be seen in the dead trees hugging the American shoreline of North Carolina. At least 10% of the forest in the Alligator National Wildlife Refuge has died within the past 35 years from salt that seeps into the soil from below or washes over the land with storm surges. Similar "ghost forests" haunt the Chesapeake Bay, New Jersey, and parts of the American Gulf coast as well (Uri 2021).

Research suggests droughts play an important role in saltwater intrusion. During times of protracted drought, the water table shrinks, and the volume of freshwater delivered by local rivers declines. This allows seawater to penetrate farther into wetland forests, eventually killing the trees. The forest is gradually transformed into saltwater marsh, a disaster for animals living in the woods. Some coastal species like North Carolina's red wolf are endangered. Other animals like raccoons, deer, and black bears are more common, but eventually all will be forced west toward the urban centers of the Carolina Piedmont.

Salinization is also a phenomenon in arid lands when evaporation from extreme heat pulls groundwater to the surface via capillary action. The water evaporates, leaving precipitated salts in the soil that prevent plants from drawing water. Farms in many parts of the world are failing because of salinization of soils due to climate change.

There's another item worth mentioning—one more trick up the alchemist's sleeve. For decades scientists have theorized that the acidification occurring in oceans must also be happening in freshwater lakes and rivers. A new study by biologists at the Ruhr University Bochum in Germany has demonstrated that this is indeed the case. Lakes are becoming more acidic. The mechanism of lake acidification is a little different from that in the oceans, but the results are the same, a lowering of the pH and increasing stress on shellfish and other calcifiers (Gies 2018).

Water Wars

It's easy to imagine a future where countries fight wars over dwindling water and other resources. Droughts, melting snowpack, and heat waves will strain international relations, especially when dams are built on rivers shared by several countries, depriving those downstream of adequate water. The Nile, Mekong, Tigris and Euphrates, and Jordan Rivers are transboundary rivers already sparking conflict over water rights. Similar political wrangling over water is happening in the United States, in particular between the seven states and two countries that draw water from the Colorado River (see figure 67).

Contentious politics concerning water rights applies to aquifers as well. A good example is the Ogallala Aquifer and the eight states that use its water.

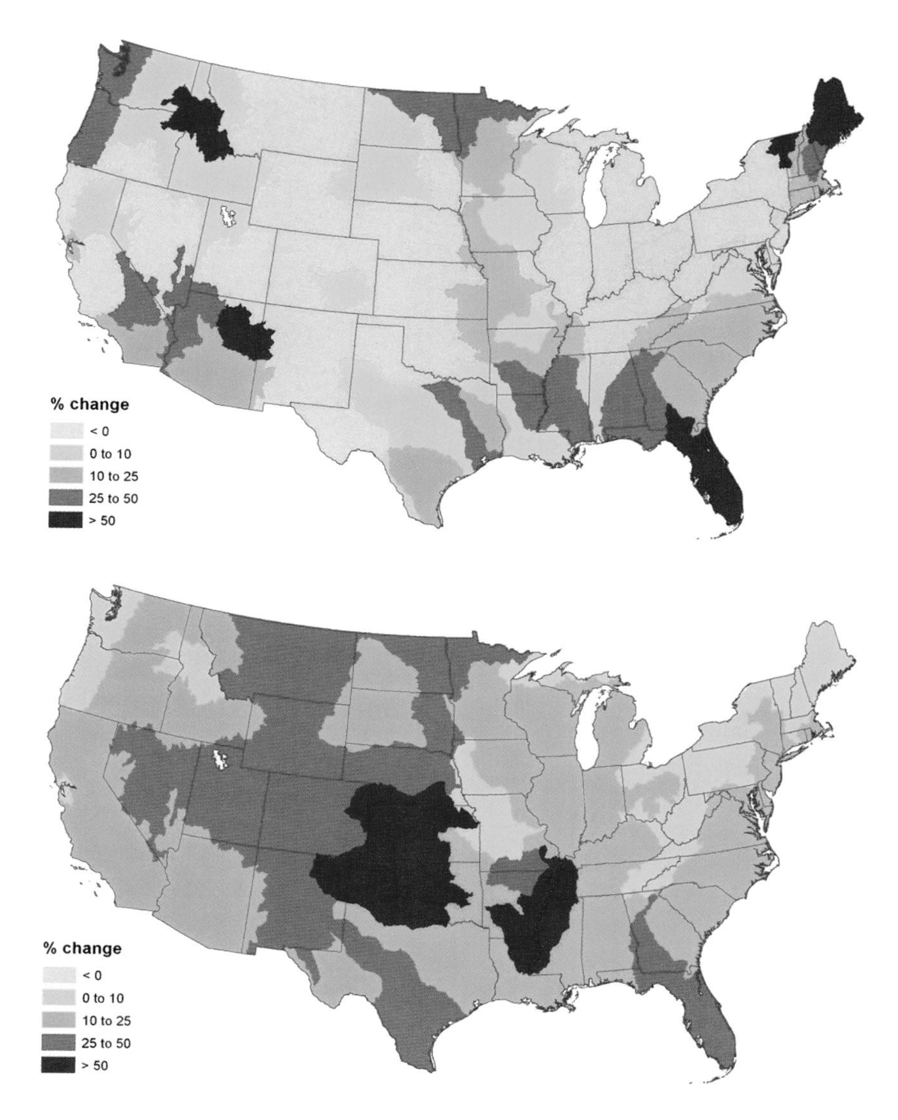

Figure 67. Maps that illustrate the projected increase in water demand between 2005 and 2060 based on US population growth and changing socioeconomic conditions. The map at the top shows increased demand for water with no climate change. The map on the bottom shows increased demand for water with climate change. Image from GlobalChange.gov, the US Global Change Research Program.

Figure 68. Ogallala Aquifer. Brown, orange, and yellow areas are where the Ogallala Aquifer is being depleted. The blue areas are where it is being replenished. If the Ogallala were to run dry, recharging it would take thousands of years (Scott 2019).

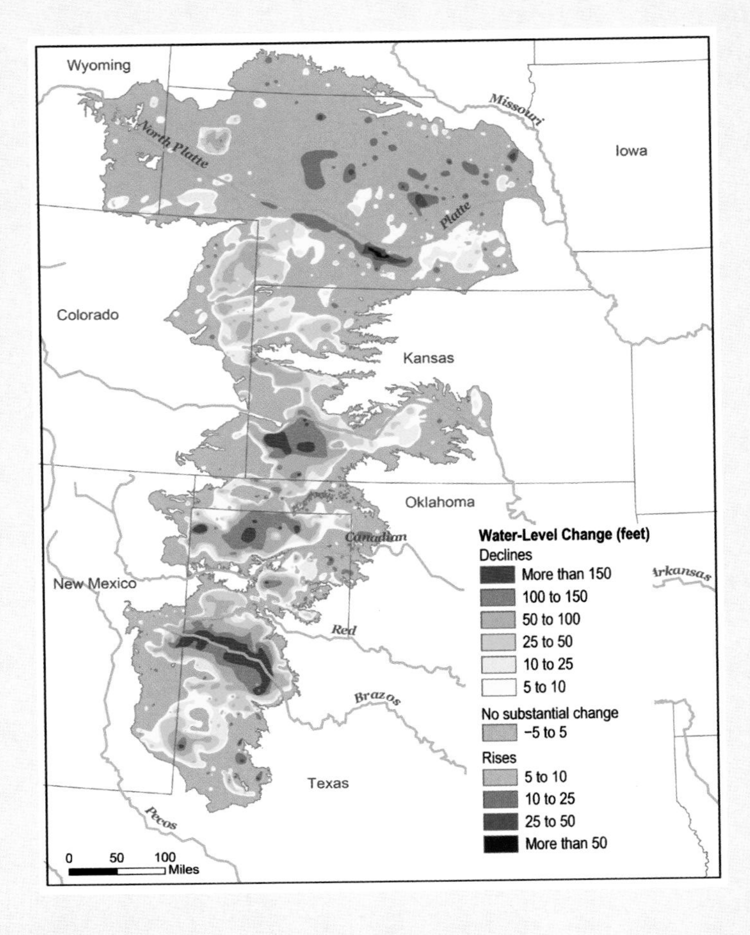

The Ogallala Aquifer

An aquifer is an underground, permeable rock formation that holds water. Much of the world's water for drinking, agriculture, and industry is pumped from aquifers. Normally, water withdrawn from an aquifer would be recharged by rain, but aquifers are disappearing globally from overextraction and increased evaporation due to climate change's higher temperatures.

The Ogallala Aquifer (the largest aquifer in the United States) supplies drinking water for 2.3 million people and sustains a fifth of the cotton, wheat, corn, and cattle grown in the United States (USDA 2018). But farming and mining interests are draining the aquifer, and agrochemicals are polluting its waters. Within 50 years, the Ogallala is predicted to be 70% gone. Overuse, climate change, and government subsidies that encourage cultivating water-intensive crops like corn (only profitable because of the subsidies) jeopardize the Ogallala's future (Sanderson, Griggs, and Miller-Klugesherz 2020).

The Ice Wars Cometh

Brewing in the still-icy waters of the Arctic is an alarming and growing potential for armed conflict. Russia is expanding military bases along its Arctic shoreline, testing nuclear weapons in the region, flying bombers into Canadian airspace, and ignoring international treaties by claiming half the Arctic Circle as Russian territory. Its navy even planted a Russian flag on the North Pole's seafloor!

China, a self-proclaimed "near Arctic Nation," lies 900 miles (1,450 km) from the Arctic Circle, yet is steadily expanding its military presence farther north, building so-called Arctic research bases, purchasing a Swedish submarine base, and engaging in naval exercises off the coast of Alaska. Chinese companies have tried (unsuccessfully) to buy an old naval base in Greenland and to purchase about 97 square miles (250 km^2) of land in Iceland to build an airstrip and golf course (Doshi, Dale-Huang, and Zhang 2021). Really? A Chinese golf course just below the Arctic Circle 5,000 miles (over 8,000 km) from Beijing?

Why the interest in the Arctic? Because shipping through an ice-free Arctic will be 40% cheaper than going through Panama. Moreover, the lure of vast, untapped, and largely unexplored Arctic mineral wealth is a siren song too alluring for nations to ignore. Whoever controls the high ground wins the battle. And there is no higher high ground (latitudinally speaking) than the Arctic.

What to Do

- **Gray water recycling.** Gray water is wastewater (excluding fecal matter) from houses, office buildings, and industry. Gray water can be recycled to free up pure water for other purposes. This will also reduce the amount of gray water discharged into the environment.
- **Water conservation.** Basically, this means use less water in home and businesses.
- **Water storage.** Increase the amount of water storage as insurance against periods of water insecurity.
- **Kill your lawn.** A lush, suburban lawn masks a multitude of ecological sins. It displaces native flora and fauna; pollutes local creeks with pesticides, herbicides, and fertilizers; causes eutrophication; and depletes city water supplies. Moreover, in most locales watering a lawn uses a fossil fuel–based electrical grid that emits carbon.

  In 2021 the state of Nevada outlawed lawns. Citizens in Las Vegas have until 2027 to replace their sprinkler-watered lawns with native vegetation. Drought and biodiversity loss have rendered the lawn aesthetic, especially as it is practiced in the United States, a hopelessly antiquated custom. Kill your lawn . . . and go native.

- **Eutrophication.** There is no easy way for societies to adapt to eutrophication of large bodies of water. The only solution is to mitigate the flow of nutrients. Typically this involves (1) planting a buffer zone of trees along streams and lakes to absorb runoff or (2) letting rivers overflow their banks, thereby using floodplains to filter nutrients. However, adaptation on a small scale is feasible. Cities can physically remove algae from ponds, add sulfates to chemically neutralize phosphorus, or (as in Charlotte, North Carolina) create artificial rapids with boulders to aerate local creeks and thus slow the effects of eutrophication.
- **Monitor aquifers.** Places like California that depend on melting snowpack for water may have to start tapping aquifers, carefully monitoring aquifer recharge/extraction levels to prevent depletion.
- **Crop switch.** Water shortages will force farmers in some areas to switch to dry-climate crops. Simple techniques like leaving the stubble of the previous season's crops in the ground can reduce evaporation of water from the soil.
- **Desalination plants** are on the rise as water shortages increase and the technology becomes more cost effective. But desalination harms marine life and is an energy-intensive operation run mainly with fossil fuels. Currently there are 20,000 desalination plants worldwide (Spiegel 2021).
- **International cooperation.** In a perverse way, the militarization of the Arctic could be seen as a logical adaptation to a new climate reality. But while nations spend billions in an ill-advised polar arms race, the sea, silent and unnoticed as a cat, creeps inland; temperatures rise; hurricanes gather force; and the climate becomes wilder and more unhinged. Cooperation, not competition, is what will keep Project Civilization going.
- **Ice stupas**, a recent innovation coming out of Kashmir, India, are essentially artificial glaciers (see figure 69) that supply water for farmers who, for centuries before the advent of climate change, watered their crops with meltwater from natural glaciers and snowpack (Kolbert 2019). As with most mountain glaciers, those in the Himalayas are rapidly melting.
- **Warka Towers.** Cheap low-tech towers designed to pull moisture out of the air through condensation (see figure 70) are the brainchild of Italian designer Arturo Vittori. The nongovernmental organization

Figure 69. Ice stupas are artificial glaciers that allow farmers to water their crops. They are constructed during winter days by bringing meltwater through a hose to lower elevations near a farmer's fields. At night the temperature drops, and the meltwater refreezes. Gradually, over many weeks, a towering ice stupa forms and then melts in the spring, providing much-needed water for farming. Photo by Vasantha Yoganathan, *New Yorker* © Condé Nast.

Figure 70. Warka Towers are cheap low-tech towers designed to pull moisture out of the air through condensation. Designed by Arturo Vittori, Warka Towers have been sucessfully employed in rural Ethiopia, Cameroon, Togo, and Haiti. Photo © Warka Water, Inc.

Warka Water's water-delivery towers were first built in Ethiopia, supplying as much as 25 gallons (95 liters) a day to poor rural villages that often had limited access to clean water (Warka Water: Every Drop Counts n.d.; Engineering for Change 2022). There's no reason such structures can't work in other countries as well.

- **Relocation.** The American Southwest faces hotter temperatures, increasing water insecurity, floods, wildfires, and other climate calamities. Will technological innovation and water conservation save the day? Should a city the size of Los Angeles continue growing in a fast-drying desert? At what point is relocation the best adaptation strategy for residents in lands with a diminishing water supply? We argue for retreating from a rising sea (see "Sea Level Rise"). Should cities retreat from an expanding desert? To stay or not to stay . . . that is the question.

SPACE

Climate Refugees

"Give me your tired, your poor,
Your huddled masses yearning to breathe free,
The wretched refuse of your teeming shore.
Send these, the homeless, tempest-tost to me,
I lift my lamp beside the golden door!"
—Emma Lazarus, "The New Colossus"

After Us, the Deluge

On February 24, 2022, Vladimir Putin launched a brutal invasion of Ukraine, razing entire cities and indiscriminately killing thousands of civilians. By the fall of 2022, 7.6 million Ukrainians (including nearly 3 million forcibly taken to Russia) had relocated to neighboring countries, commencing Europe's largest refugee crisis since World War II. Unlike the refugees from the 2011 Syrian civil war, the Ukrainians were mostly welcomed in their host nations, in part because of shared cultural affinities with other European countries but also from a mutual recognition of the danger posed by an expansionist Russia.

While 7.6 million may seem like a large number, the United Nations predicts 200 million refugees will be seeking sanctuary within or beyond their respective nations by 2050 because of climate change (Gaynor 2020). The Institute for Economics and Peace (IEP), an international think tank that measures threats to peace, puts the number of 2050 climate refugees considerably higher, at over 1 billion (Baker 2020). Another study by Cornell researchers contends that by century's end 2 billion people (one-fifth of the world's population) will be climate refugees just from rising seas. Added to those numbers will be a deluge of tens of millions of people fleeing droughts, floods, wildfires, storms, food shortages, and a host of other climate calamities spawned by an overheated atmosphere (Cornell University 2017). Even if the IEP figure is inflated, it's clear that humanity faces an unprecedented refugee crisis on a planetary scale.

What Is a Refugee?

Refugees are people who have been forced to leave their country or community in order to avoid violence, persecution, or a natural disaster and who cannot return home safely. Those who flee primarily because of climate change are referred to as *climate refugees*, though in fact changing weather patterns are often one of several interacting factors contributing to migration.

Climate refugees are not mentioned in the 1951 UN Refugee Convention, which defines refugees within the narrow context of persons fleeing across national borders from political persecution in its various guises. People forced to relocate due to extreme weather events have no legal status as refugees and therefore scant hope of finding asylum, unless some other factor contributes to their status (Goodwin-Gill and McAdam 2017). This was the case in Syria, where crop failures from groundwater depletion and drought (probably enhanced by climate change) led to political unrest and civil war. Nearly 7 million Syrians were granted refugee status because of persecution from President Bashar al-Assad and his Russian allies but not because of any purported connections to a changing climate.

The Clash of Ignorant Armies

Many see the war in Ukraine as a throwback to the 20th century's most barbarous episodes, another depressing reminder of the inhumanity of humanity. But there are implications to the conflict that go beyond the suffering of the Ukrainian people and the apparent inability of our species to override its native aggressions. Russia is the world's fourth-biggest greenhouse gas emitter and a major exporter of oil, gas, and coal. We (the citizens of the world) need the cooperation of the Russian government (and the Chinese government) to curtail global carbon emissions. Geopolitical conflicts are distractions that prevent us from dealing with climate change in a timely fashion. They also add to our climate woes.

If the US military were a country, it would rank as the 47th-largest emitter of greenhouse gases. An estimated 6% of humanity's annual carbon emissions come from the world's armed forces, a figure that does not include emissions discharged during armed conflict. The Gulf War (Desert Storm) lasted only 43 days yet produced 2%–3% of that year's global greenhouse gas emissions, in large part because Saddam Hussein in an act of environmental terrorism set fire to Kuwait's oil wells. The current military buildup of China and Iran and the ensuing arms race with the United States will exacerbate our climate problems. Carbon emissions will rise with the expansion of bases, military exercises, and the production of weapons (and of course during war), making it all but impossible for nations to achieve carbon neutrality (McCarthy 2022).

As the climate deteriorates, competition between and within nations over dwindling resources will increase. Russia, the United States, Canada, and China will compete for access to an ice-free Northwest Passage. Water wars are likely to break out in the Middle East and elsewhere. A hotter climate means crop failures, economic stress, and more refugees, increasing the potential for open war, which not only discharges more carbon but also disrupts global food supply chains. We learned this lesson from the war in Ukraine, where exports of Ukrainian grain to African countries were blocked by Russia, exacerbating famine already underway in parts of East Africa (see "Famine" in "Earth"). Global warming, famine, war, more warming... if you're not careful, the many feedback loops of climate change will make your head spin.

Modern warfare is a savage affair, unleashing powerful weapons that inflict pain on any life that gets in harm's way, human or otherwise. Its savagery is evident in photographs of the American Civil War, where scattered among the bodies of fallen soldiers and the bloated carcasses of livestock lie shattered trees and even entire forests decapitated by artillery. Agent Orange devastating people and wildlife in Vietnam; tigers stepping on land mines in Cambodia; toxic chemicals from bombed factories polluting the air, land, and groundwater of Ukraine... we rarely recognize the impact of war on the environment. Genocide is considered by the UN to be a crime. Ecocide is not. When we tally the numbers of refugees, we should include those fleeing on four legs (and consider the damage to the biota that can't flee because they are rooted to the ground). If we are to ever make peace with the climate, we must also make peace among ourselves.

Diaspora

Climate change conceals a terrible environmental injustice. The world's most marginalized and disenfranchised, the poorest in Africa, South and Southeast Asia, Latin America, and elsewhere, have contributed the least to global warming, yet are the most affected. Migration is a logical adaptation to the impacts of a warming world, but for hundreds of millions living in poverty, leaving one's home to move to another country is an extreme economic and psychological hardship. Many will become internally displaced within their own country. Others will shelter in place and tough it out for as long as possible. Eventually most will try to follow the mass exodus of their fellow citizens, paying the migration smugglers what they can to join the climate diaspora.

Climatologists predict that life for anyone between 45 degrees latitude north and 45 degrees latitude south will become increasingly difficult (if nothing is done to mitigate climate change) (Wallace-Wells 2020). In many countries, temperatures will become so hot and humid as to be unlivable. Working outdoors will be a death sentence. Heat coupled with drought, food shortages, and economic collapse will force millions to move elsewhere. Like plants and animals (see "Nature on the Move"), humans will move upslope and toward the poles to escape the tribulations of a warming world. The vast majority will migrate north. Why north? Because that's where the lion's share of Earth's climate-resistant real estate will be. It's a matter of geographic happenstance.

As figure 71 illustrates, there is almost no land in the Southern Hemisphere south of 45 degrees latitude. Outside of some mountainous areas with sufficient rainfall, the continents of Africa and Australia (except for Tasmania) and most of South America are within the no-man's-land of climate change. Refugees from there will have little choice but to move north toward the relatively cool and economically prosperous countries of North America and Europe.

The Promised Land

The Great Climate Migration has already begun. Californians are leaving the state to escape mudslides, drought, and wildfires. Pacific Islanders, people on the Mississippi Delta, and coastal Native Alaskans are moving entire communities away from the rising sea. The so-called immigrant caravans from El Salvador, Nicaragua, and Honduras are in many cases farmers fleeing a climate-change-induced drought or else the devastation of powerful hurricanes. Facing starvation after their crops failed, the farmers and their families migrated to overcrowded cities where jobs were scarce and gangs

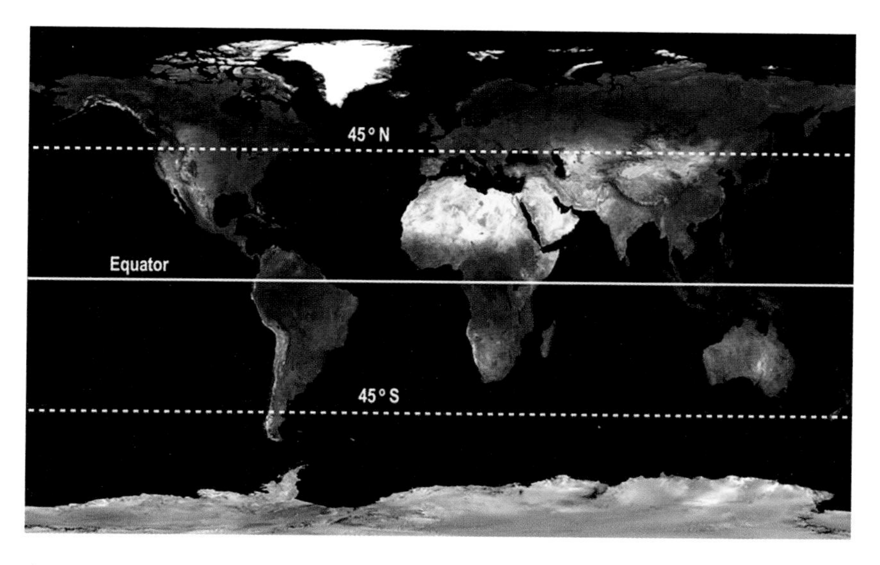

Figure 71. In the coming centuries, hundreds of millions of people will flee the overheated lands between latitudes 45 degrees north and 45 degrees south. Because there are more land, jobs, and political stability in the temperate regions of the Northern Hemisphere, most climate refugees will migrate north to the United States, Canada, Europe, and likely Russia as well. There is nowhere to go to escape climate change in the Southern Hemisphere (except perhaps to higher elevations). Image from NASA, modified by Charles Pilkey.

many. From the desperate cities they walked hundreds of miles north to the Promised Land, rumored to lie just beyond the Mexican border. A similar migration pattern from hinterland to city to climate sanctuary in the north is unfolding in other parts of the world.

But the climate is changing everywhere. There is no "Promised Land," and the phrase *climate sanctuary* (or *haven*) is a relative term. Countries in the far north are relatively better off than those in the far south, at least in terms of average temperatures, GDP, political stability, and job opportunities. Yet Canada, the United States, Europe, and Russia are suffering through their own array of climate disasters—wildfires, heat waves, floods, and the like. And they will witness their own share of internally displaced climate refugees. Florida alone is expected to produce 5 million refugees as rising seas overwhelm its coastal communities.

Today's trickle will be tomorrow's flood. What will happen when millions of climate refugees pile up on our borders? Will we greet the huddled masses yearning to breathe cool air with the same compassion given Ukrainian refugees, or will we abandon them in cramped refugee camps? Knowing the new migrants are likely to be culturally and ethnically different from us, will we, the "privileged" children of Western democracies, welcome them with open arms or with closed borders? How far can compassion be stretched in a world with a climate gone mad?

Watch this space . . .

What to Do

- **Refugee recognition.** The UN must officially grant refugee status to those forced to immigrate because of climate-change-related events.
- **Refugee planning.** As outlined in the Biden administration's 2021 report on climate migration, we should do the following:

    1. Set up an early warning system that monitors the potential for widespread climate migrations before they occur.
    2. Increase US foreign aid to help people adapt to climate change so they can stay in their respective countries for as long as safety allows. This includes teaching farmers climate-smart agricultural practices.
    3. Plan for the coming deluge of climate refugees by constructing humane refugee shelters, finding communities to sponsor refugees, increasing staff to aid asylum seekers, and implementing an education (including job training) and health system to integrate refugees into society.
    4. Educate communities to counter the inevitable xenophobic, nationalistic, and racist response to large numbers of immigrants.
    5. Work with other governments to reduce human trafficking, illegal mining and resource extraction, and economic activities directly related to migration.

- **Reduce armed conflicts.** We need to stop waging pointless wars—wars that emit unwanted CO_2, increase the numbers of refugees, and cause unnecessary suffering for people and for Nature. If we can't stop wars entirely, then we should try to reduce their occurrence. Promoting democracy and a free press with high regard for empirical truth would be a good start toward countering the lies and propaganda of authoritarian leaders who might otherwise lead a country into war.

Climate Havens

There's no place like home.—L. Frank Baum, *The Wonderful Wizard of Oz*

What Makes a Climate Haven?

Our home is our haven. We celebrate our homes because they protect us from whatever threatens us from outside. Weather events, bad actors, and wild animals are why our species has always sought shelter. Now our most treasured havens are endangered from climate change. As sea level rises and wildfires burn, what will our future havens look like, and where will we be making them?

Climate havens are locations where the worst effects of climate-induced disasters are minimal, and political will and infrastructure exist to handle future climate issues. As extreme weather events worsen over time, more people will consider relocating to safer places. Climate relocation is different from climate migration. Climate migration means moving after climate change has made life untenable. Climate relocation is planning your life to voluntarily move to a climate haven.

A climate haven includes the following: fewer severe climate shocks, reliable access to fresh water, high vacancy rates and/or affordable housing, robust infrastructure, a thriving job market, access to cheap, local crop yields, good health care facilities, and a proactive government that has plans and resources to cope with climate change. Climate havens can better withstand the most devastating impacts of the climate crisis, but they will still be affected by extreme weather events along with the rest of the planet. You can run, but you can't hide from a changing climate, at least not indefinitely.

Rules for Locating a Climate Haven

- **Get away from the coast** (including areas affected by hurricanes). Assume the worst sea level rise predicted by models is going to happen.
- **Get away from wildfires** (including wildfire smoke pollution). This is a hard one since smoke pollution can travel great distances.
- **Other things to avoid.** Tornadoes, bad air quality, and places with a high percentage of properties in flood zones.
- **Go north and/or upslope** (toward cooler temperatures).
- **Find places with reliable freshwater sources.** Lakes, rivers, creeks, springs, aquifers, and reliable precipitation (hard to predict with current climate models).
- **Seek infrastructure.** Stable governments that have plans to deal with climate change, medical facilities, dependable energy sources, complex food distribution networks, firefighting teams, physical security (including police and/or military), good roads, and so on.
- **Seek towns with . . .** cheap housing, robust economies, and employment opportunities; also, towns that welcome migrants.
- **Move early** to avoid the rush and the inevitable backlash against climate refugees (the "Floridafication" of northern cities or "Californication" of the Pacific Northwest).
- **Keep in mind:** the rules for locating a climate haven are provisional and may change with changing weather patterns. Once at the top of most climate haven lists, Seattle, Washington, and Vancouver, Canada, have slipped a notch or two because of recent wildfires, heat domes, and massive floods. Today's climate haven may become tomorrow's climate hell.

The Hot Zone

Most climate scientists say land between 45 degrees latitude north and 45 degrees south will eventually become too hot for people to live (Wallace-Wells 2020). There are exceptions to this rule. New Zealand's North Island

and Tasmania are in climate change's "hot zone" yet are considered to be climate havens. Heat waves on both islands are moderated to some extent by the mountains and the sea, though coastal flooding, droughts, and wildfires are expected to increase in the coming years.

Africa, southern Asia, and all of Latin America (except Patagonia) lie within the hot zone. All are in trouble. It is from these densely populated regions that the majority of climate refugees will depart, many seeking asylum in the United States, Canada, Europe, and probably Russia (see figure 72).

Their diaspora will deliver what will be the largest mass migration of humans in our species' history. Billions will flee rising seas, droughts, fires, famine, and heat waves. Most, at least initially, will relocate within their own national borders. Others will move to neighboring countries. Millions will make the long journey to the far north, their exact numbers determined by how fast the climate destabilizes and how quickly the world's governments can reduce emissions.

Shelter from the Storm

Climate havens in the United States are generally clustered in three regions: New England, the Pacific Northwest (including Alaska), and the Upper Midwest. New England's havens are found in upstate New York (along Lake Erie and Lake Ontario), Vermont, New Hampshire, and parts of Massachusetts and Maine (away from the coast). The Pacific Northwest, as discussed above, has lost some of its attraction as a refuge because of fires and wild weather events, though there may still be scattered pockets of refugia along the corridor from Northern California to Canada. Some sources cite eastern Washington, Idaho, western Montana, and Appalachia as viable climate havens, although the wildfires of tomorrow may have something to say about those assessments.

At this time, the Upper Midwest, especially along the Great Lakes, from Minnesota to western Pennsylvania, may well be America's best climate haven. Flooding is an issue there, but so far the region has avoided the catastrophic wildfires (though not the wildfire smoke) and heat waves of the West (knock on wood). Ample water, local food sources, no rising seas, cities with climate change plans in the works, and cheap housing from years of depopulation after American industries went overseas qualify Midwestern cities as climate havens. Some cities like Buffalo, New York, and "climate-proof Duluth" (Minnesota) are actively promoting their communities as climate havens, hoping new arrivals will boost the local economy.

Tulane University associate professor Jesse Keenan provided a list of potential climate havens in the United States, most of which are located in the

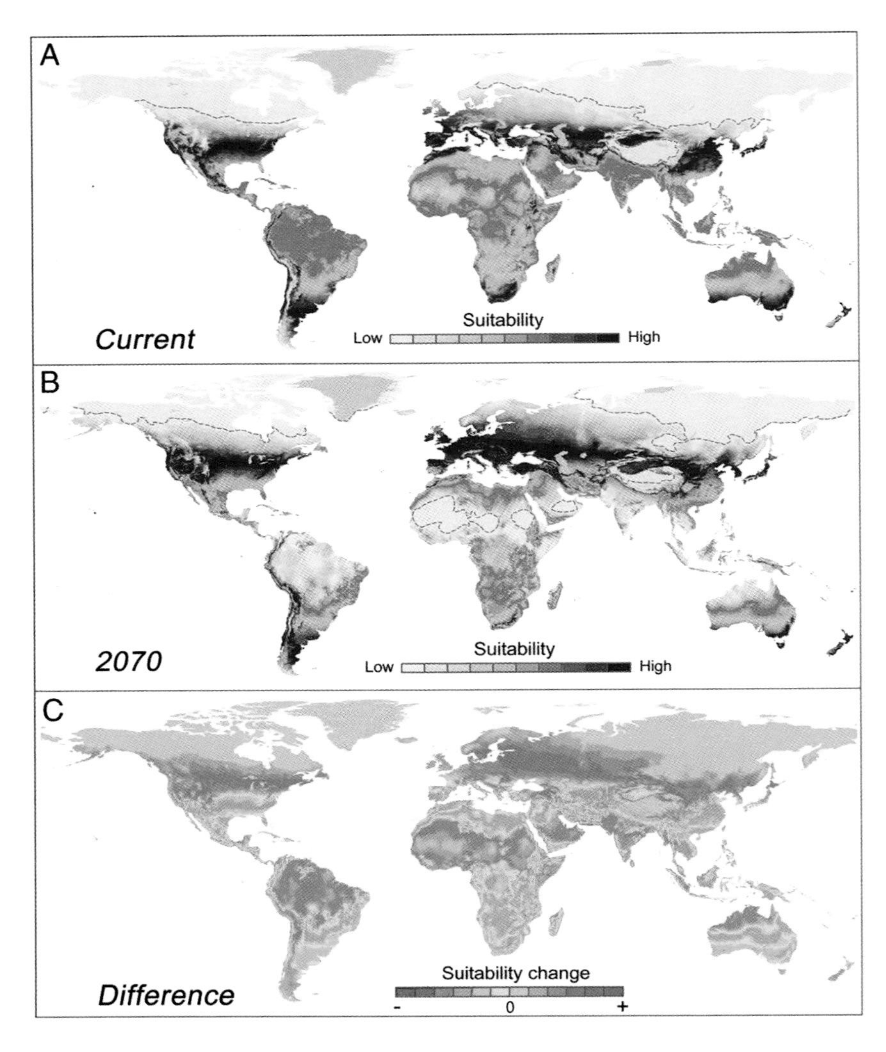

Figure 72. The three world maps illustrate where climate is most suitable for human habitation. Map A shows the situation in 2020. Map B, for the year 2070, illustrates a significant northward shift in suitability as the climate warms. The dark areas on the top two maps are the most suitable. Note that there are almost no suitable areas in the Southern Hemisphere, either at present or in 2070. Map C illustrates the degree of change in suitability between the present and 2070. Image from Chi Xu et al. (2020).

East and Midwest. Keenan stressed that these cities could take steps today to make it easier for them to welcome climate migrants in the future. "This isn't we're going to build a community for tomorrow," he told CNBC. "We're going to build a community for today. And that's going to be the foundation for the building of a community for tomorrow" (quoted in Jacobson 2022).

Table 13 lists North American cities deemed climate havens by Keenan and others. (The list is in alphabetical order, not in order of preference as climate havens.) The cities on the list will almost certainly change as unforeseen climate disasters unfold. But for the next 30 years or so, these (we think) will be relatively secure shelters from the coming storm called climate change. After 2050 all bets are off. We invite the reader to research which climate haven is best for their particular needs. Table 14 ranks 20 countries from least to most vulnerable and readiness to adapt to climate change.

What to Do

- **When is it time to relocate?** This is a highly personal question that touches on socioeconomic and psychological factors. For many families, relocation is not economically feasible. Moreover, some may have deep historical/emotional bonds to a particular location, family members with health issues, job security, children in school, and the like. Staying in place may be preferable to the stress of relocating to some abstract climate haven. However . . . if your home is in imminent danger from floods, wildfires, storms, or sea level rise, you should have left yesterday.
- **Make shelters for refugees.** In the months after Hurricane Maria devastated Puerto Rico, an estimated 300,000 Puerto Ricans fled to the US mainland, overwhelming FEMA (the Federal Emergency Management Agency), the Red Cross, and other volunteer organizations around the country. It's clear the United States doesn't have the systems in place to feed, shelter, and assimilate the tens of millions of climate refugees that will arrive from abroad or those that will relocate from within our national borders in the decades to come (see "Climate Refugees"). Federal, state, and local governments need to start planning now for the approaching deluge.
- **Affordable housing for refugees.** This is a global problem. The flood of refugees pouring across the United States later this century will use up the affordable homes and low-rent apartments in climate havens like the Midwest. We need to find or make cheap housing to accommodate a growing refugee population. Some possibilities are:

  Refurbish old structures. Defunct warehouses, factories, or stores can be repurposed as apartment buildings, community centers, retirement homes, or refugee centers. This is an old idea

Table 13. Climate Haven Cities: United States and Canada

Ann Arbor, Michigan	Cleveland, Ohio	Milwaukee, Wisconsin
Asheville, North Carolina	Denver, Colorado	Minneapolis, Minnesota
Bangor, Maine	Detroit, Michigan	Pittsburgh, Pennsylvania
Boise, Idaho	Duluth, Minnesota	Rochester, New York
Boulder, Colorado	Erie, Pennsylvania	Spokane, Washington
Buffalo, New York	Grand Rapids, Michigan	Thunder Bay, Ontario
Burlington, Vermont	Green Bay, Wisconsin	Toledo, Ohio
Chicago, Illinois	Madison, Wisconsin	Quebec, Quebec

Sources: Keenan in Jacobson (2022); Brentjens (2022); Pogue (2021).

Table 14. Notre Dame Global Adaptation Initiative "Top 20" Index of Countries (2021)

1. Norway	11. Austria
2. Finland	12. Australia
3. Switzerland	13. Luxembourg
4. Denmark	14. Canada
5. Singapore	15. Republic of Korea
6. Sweden	16. France
7. Iceland	17. United States
8. New Zealand	18. Netherlands
9. Germany	19. Japan
10. United Kingdom	20. Ireland

Note: These 20 countries (ranked from least to most vulnerable) are considered to be the least vulnerable to climate change. Vulnerability is a measure of a country's exposure to the negative effects of climate change as well as its capacity to adapt to those effects. Access to food, water, health care, housing, and infrastructure are factors that define climate adaptability. Norway is at or near the top of most lists of climate-resilient places while some countries in sub-Saharan Africa and the Pacific atoll nations (not shown here) are generally at the bottom of such lists. Because the world's climate is unfolding in ways that are hard to predict, the "Top 20" list may look very different in the decades to come. Source: Notre Dame Global Adaptation Initiative, 2023, https://gain.nd.edu/our-work/country-index/rankings/.

in the southern United States, where cities have renovated tobacco warehouses and textile mills, turning them into shopping malls and apartments.

3D printing. This is a cheap, energy-efficient way to quickly make houses in concrete. The process emits less carbon dioxide (CO_2) than conventional building approaches. Will 3D-printed refugee camps one day be employed as a humane alternative to housing refugees in tents?

Modular units. Buildings in wood or steel, stacked like Lego blocks, can be constructed off-site quickly and with less waste than homes built using current on-site methods. Because modular units are transportable, they can be easily moved should sea level rise, wildfires, or other agents of climate change make such a move necessary.

Tiny houses. This resurgent American architectural and social movement stands in stark opposition to the standardized McMansions typical of modern US suburban neighborhoods. The movement, which has spread to Europe, advocates living in affordable, downsized (typically less than 500-square-foot /~46 m^2), environmentally friendly houses. Sometimes tiny houses are on wheels, a useful feature for escaping extreme weather events.

Container homes. Some Americans have converted shipping containers into houses, sometimes combining two or more containers to create more living space. Especially popular in Alaska, the price of a container house can be as low as $15,000 plus the cost of renovations (windows, doors, insulation, plumbing, etc.). For the average middle-class American, the prospect of living in a metal box is a bad joke. For those fleeing war and ecological ruin in the overheated south, desperate enough to seek sanctuary by crossing jungles, deserts and even oceans, such a dwelling is a veritable palace. One person's box . . . another person's castle.

Green Cities

Hot town, summer in the city
Back of my neck gettin' dirty and gritty,
Been down, isn't it a pity?
Doesn't seem to be a shadow in the city.
—John Sebastian, Mark Sebastian, and Steve Boone,
"Summer in the City"

Cities and Climate Change

The city . . . center of commerce, transport, religion, and government . . . site of the world's finest museums, sports arenas, restaurants, and concert halls . . . the beating heart of modern life. Today's conurbations in steel and glass may appear vibrant, robust, and eternal. But the archaeological record is chock-full of defunct civilizations crumbled and rotting beneath the ground, casualties of war, economic decline, and collapsed ecosystems. The city is in reality a fragile enterprise.

With more than half the world's population (increasing to 68% by 2050), cities have become major sources of carbon emissions, discharging about 70% of the world's total greenhouse gases. As our climate continues to unravel, urban life will become more challenging, and in some places, untenable. Rising seas will overrun coastal towns. Heat waves and droughts will bake desert cities to the point where they are unlivable. Failing economies will unleash a desperate diaspora as millions of urban poor stream out of the

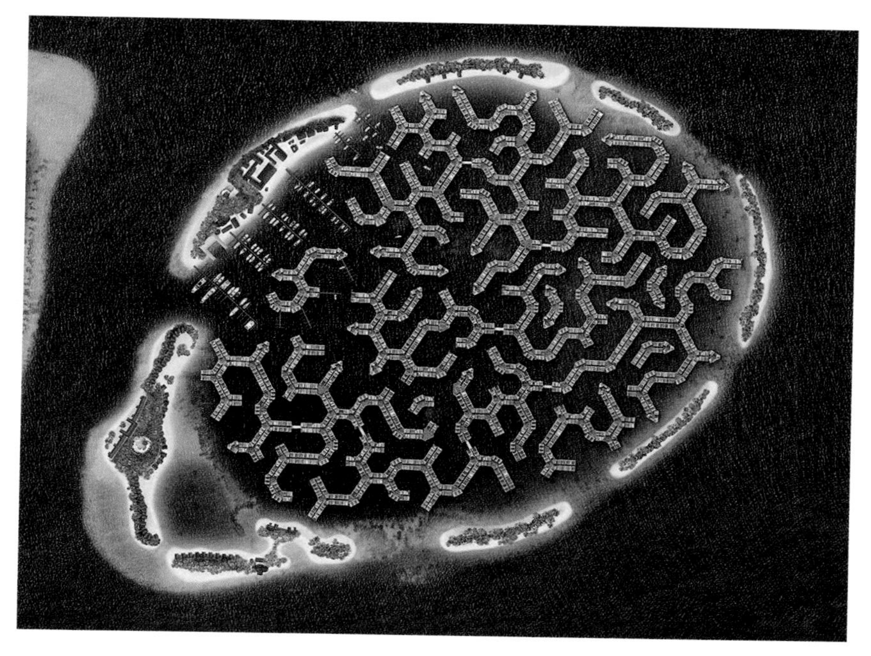

Figure 73. Rendering of a planned floating city in the Maldives. A joint venture between Dutch Docklands and the Maldives government, the floating city is an eco-friendly settlement of several thousand floating houses on a lagoon near the capital, Male. It is designed to accommodate 100 years of sea level rise. Image © Architect: Koen Olthuis, Waterstudio.NL. Developer: Dutch Docklands.

tropics seeking sanctuary in the cooler (relatively speaking) northernmost cities of Europe and North America (see "Climate Havens").

In a climate grown hot, unpredictable, and increasingly hostile, we need to redesign urban areas to make them resilient to climate disruptions. The good news is that many of our most creative engineers, architects, and city planners are doing just that, conceiving strategies whereby urban populations can survive global change. Some of their proposals are rooted in ancient practices thousands of years old. Some are based on new technologies. And some at first glance may seem a little wacky yet in retrospect turn out to be eminently practical, producing a "Well of course, why didn't we think of that before?" moment (see figure 73).

Most cities have drafted plans for climate change mitigation but not for adaptation. City planners need to devise ways to cope with their community's unique set of climate issues. What can be done to help low-income families survive extreme heat, floods, and sea level rise? How should new buildings be designed for tomorrow's climate? What happens when the reservoirs run dry or when the rains fall incessantly?

The Urban Heat Island Effect

Cities are generally hotter than the surrounding countryside. That's because buildings, roofs, and streets absorb and re-emit more heat from the sun than do natural surfaces. Urban heat also comes from industry, car exhausts, air-conditioning, electric lights, sunlight reflected from windows, and even the swelling masses of humanity itself. According to the US Environmental Protection Agency, cities can be as much as 7°F (4°C) hotter than adjacent rural areas, enough of a heat differential to change local weather patterns (EPA 2022). When coupled with a rapidly warming planet, the urban heat island effect will make life in many cities unbearable, especially for the poor who can't afford air-conditioning.

What to Do

HEAT AND DROUGHT

- **Shade.** Awnings, pergolas, shutters, sunscreens, and other architectural shading can offer relief from heat waves and ameliorate the heat island effect. But the simplest way to cool a city is to plant trees. Urban forests provide shade, reduce runoff, filter air pollutants, create wildlife habitat, pull carbon from the air, and cool the atmosphere by evaporating water (evapotranspiration). Ideally those native species better suited for surviving the hotter climes of the future should be planted preferentially. Singapore is an example of a city planting shade trees everywhere, including on the sides and roofs of buildings. We must be wary, however. Enhanced tree cover in some areas may increase the risk of urban wildfires.
- **Stop urban deforestation.** This is a corollary to the analysis presented above. We need to find ways to house our growing urban populations without razing entire forests and killing every animal that lives there. A strictly enforced ban on urban deforestation will compel developers to find ways to construct housing developments in a way that minimizes damage to the forest ecosystem. Another approach might be to increase the number of forested parks in a city and connect those parks via wildlife corridors to other forests. Developers . . . spare that tree!
- **Water parks.** When drought is not an issue, then free public splash parks, swimming pools, and drinking fountains can offer welcome and potentially life-saving relief from heat waves (especially for the urban poor).

Figure 74. Uchisar Castle, Cappadocia, Turkey. At one time about 1,000 people lived inside this structure carved from a rock outcrop. Could a similar approach be used in parts of the American Southwest to cope with rising temperatures? Photo by Morphosis, https://commons.wikimedia.org/wiki/File:Uchisar_Castle,_Nevsehir.jpg.

- **White cities.** Painting rooftops (and walls) white reflects more sunlight (albedo effect) than conventional roofs, potentially cooling a roof by as much as 60° F (33° C). In addition, emerging technologies like clear binders can be added to asphalt to increase its reflectivity.
- **Building orientation.** New buildings should be oriented in a way to minimize exposure to the afternoon sun and maximize cooling from wind. The longest side of a building should never face west. In hot climates the main room(s) should face north (south in cold climates). Buildings, streets, and even trees can be positioned to channel the prevailing winds for cooling purposes.
- **Underground cities.** People have been living underground to avoid the heat (and invading armies) for thousands of years. Examples include our cave-dwelling ancestors, underground cities like those in Cappadocia, Turkey, and more recently underground shopping malls and parking lots, often clustered around subway stations. Living underground with proper ventilation offers protection from heat waves, wildfires, and extreme weather events while freeing more land for Nature and for agriculture (see figure 74).

- **Retrofitted buildings.** Houses, apartments, skyscrapers, shopping centers, and the like can be remodeled to make them more energy efficient, to reduce their carbon footprints, and to make them more resilient to wildfires, storms, and other depredations wrought by climate change. The construction industry's habit of demolition and reconstruction is an outdated energy-intensive system that releases excessive greenhouse gas emissions.
- **Sky gardens.** Tokyo, the world's most populous city, passed a law in 2001 requiring all new buildings to incorporate rooftop gardens. Other cities have followed Tokyo's lead. The advantages gained by rooftop greenery are cooler buildings, lower energy costs, reduced noise levels, a diminished heat island effect, increased recreational green space, and the possibility of growing food for the masses. One can envision future cities where orchards, community vegetable gardens, chicken coops, and greenhouses share roof space with solar panels and wind turbines.
- **Remote work.** Millions worked from home during the COVID-19 pandemic. In the coming decades, heat waves, dust storms, and extreme weather will revive the popularity of working and studying from home. Those working outdoors will need to shift their work schedule to the cooler morning and evening hours.
- **Reductions in water use.** Municipalities should invest time and money to fix leaking pipes and to construct systems to store rainwater and storm runoff in anticipation of future water shortages. Fines can be levied on households and industry for excessive water use during periods of extreme drought. Local, state, and national governments need to monitor depletion of aquifers and other sources of drinking and irrigation water.
- **Sharing water.** Regional and international cooperation between cities is the best way to maintain an adequate supply of drinking water for all involved. For multiple countries using the same river for drinking and irrigation (the Nile, the Tigris, and the Mekong, for example), international cooperation is a must, without which the inevitable droughts will lead to inevitable water wars.
- **Traditional approaches.** We can learn much by studying how cultures from the past adapted to hot climates. Traditional houses in the Deep South have high ceilings with ceiling fans to counter muggy summers. In ancient Babylon thick adobe brick walls kept buildings cool in summer and warm in winter. For millennia, buildings known as *wind towers* or *wind catchers* have been constructed in the Middle East to cool buildings by capturing the wind (see figures 75 and 76).

Figure 75. A wind tower or wind catcher like this one in Doha, Qatar, is an architectural feature designed to cool a building by pulling the wind from above and circulating it into the interior. Photo by Diego Delso, delso.photo. (Creative Commons Attribution-ShareAlike 4.0 International [CC BY-SA 4.0]).

Narrow streets between high buildings channel wind in the same way. Lattice windows (for wind circulation), interior courtyards with orchards, and light-colored building materials are traditional design choices common to North Africa, the Arabian Peninsula, and South Asia for coping with heat. In the Mediterranean entire cities are painted white to reflect heat. And so on. . . .

- **Natural lawns.** In an age of rampant climate change and biodiversity loss, American-style lawns are an environmental absurdity requiring near-constant watering that depletes precious drinking water. Fertilizers, pesticides, and herbicides flowing downhill from suburban yards into creeks spoil riparian ecosystems. Insecticides eliminate food sources for birds and other wildlife. Fortunately, there are many alternatives to a traditional lawn. Rock and sand landscaping with native succulents and cacti should be encouraged in drought-ridden areas like the American Southwest. Native groundcovers, mosses, ferns, herbs, shrubs, and trees can be aesthetically pleasing,

Figure 76. A simplified diagram of wind movement within a wind tower. The air circulates through various rooms, eventually exiting from another opening. Drawing by Charles Pilkey.

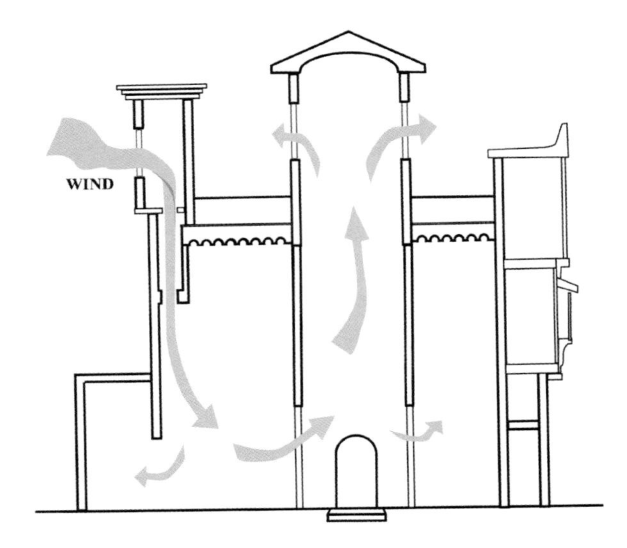

ecologically benign substitutes for chemical-fed lawns in lands with sufficient rainfall. We should perhaps rethink the composition of greens on the golf course. Artificial turf for sports fields?

- **Solar canals.** Both India and California have launched programs to install solar panels perched above irrigation canals (see figure 77). Solar canals reduce evaporation, increase available electricity, and don't use precious land. Solar panels over parking lots make perfect sense as a way to create shade *and* produce electricity.

RISING SEAS

- **Retreat.** Raising buildings on stilts (New Jersey) and elevating roads (Miami) are short-term strategies to alleviate coastal flooding. But the only permanent adaptation to sea level rise is to move away from the coast to higher ground and for government to prohibit future coastal construction in floodable locations.
- **Floating cities.** On the shores of Cambodia's Tonlé Sap, the largest freshwater lake in Southeast Asia, are floating villages that migrate with the advancing or retreating lakefront as water levels respond to seasonal changes in rainfall. Similar floating villages can be found in estuaries and harbors throughout Southeast Asia, in parts of Africa, in Lake Titicaca in Peru, and in the houseboat communities of Seattle. It's a simple evolution from a single floating house to a neighborhood of floating houses to floating cities like those proposed by the Maldives government or Oceanix, a visionary company

Figure 77. Artist's rendering of a solar canal in California. Solar panels placed above canals will reduce evaporation from irrigation and drinking-water canals. The solution simultaneously provides green electricity while preserving precious farmland. Could solar panels also be installed on pilings or on floating platforms over reservoirs to reduce evaporation and provide power without harming local ecosystems? Photo from Solar Aquagrid LLC, CC BY-ND (Attribution-NoDerivs 3.0 Unported [CC BY-ND 3.0]).

designing an affordable, sustainable floating city for Busan, Korea (see figures 78 and 79). The beauty of a floating city is that it rises with rising seas, more or less immune to sea level rise as long as it's protected from damaging waves.

- **House of warship.** In 2021 the decommissioned aircraft carrier *Kitty Hawk*, the once proud home of 5,600 sailors, was sold to a scrapyard (for one cent!). In an age of accelerating sea level rise, there are better uses for old ships than mothballing them to a scrapyard. Why not refit a ship like the *Kitty Hawk* as a "floating city" docked to the mainland in an estuary, protected from hurricanes and indifferent to rising seas? Apartments, low-income housing, condominiums, hotels, shopping centers, homeless shelters, refugee centers, schools, flight deck gardens, office space, museums, restaurants — the possibilities for retrofitting and reusing scrapped naval vessels are limited only by imagination.
- **Floating farms.** A functional floating dairy farm using electricity from a floating solar plant is now operating in Holland. The cows are

Figures 78 & 79. Two views of a floating city designed by the architecture firm BIG (Bjarke Ingels Group) in collaboration with the company Oceanix, the UN Human Settlements Program, and MIT's Center for Ocean Engineering. Oceanix Busan, the first of a planned series of floating cities, will be built off South Korea and is intended to house 10,000 people, providing energy from wind turbines and solar panels and food from conventional gardens as well as from floating reefs of seaweed, clams, oysters, and scallops. Photo courtesy of Oceanix / BIG-Bjarke Ingels Group.

fed grass clippings from nearby soccer fields and leftover vegetables from local restaurants.

- **Tidal gates.** Venice, Rotterdam, and London are cities that have built gates that can be closed to prevent flooding caused by storm surges, king tides, heavy rains, and strong onshore winds. Sea level rise will eventually make such structures worthless as waters flow around and over them.

- **The Venice solution.** Miami will not easily survive sea level rise because the porous limestone on which it is built permits seawater to seep into the city from below, making seawalls more or less useless. Some have proposed abandoning the lower stories of the city's high rises and turning the streets into canals, something like a Venice on the open ocean. The viability of such an approach requires making the submerged lower stories impermeable to seawater, constructing barriers to prevent storm wave damage, elevating roads, and rerouting electrical wires, sewage pipes, and pipes for drinking water. Is such a solution economically viable over the long term? Perhaps one day we'll find out.
- **Nature-based solutions.** Restoring saltwater marshes and mangroves can protect coastal towns from hurricane winds and storm surges. Some, like New York City, are building underwater breakwaters (living shorelines) of stone populated with oyster beds to reduce beach erosion. Nature-based approaches are temporary expedients that protect beach towns from storms and provide habitat for marine organisms. But in the long run they will not prevent sea level rise. The Staten Island project, designed for an 18-inch (46 cm) rise in sea level, will be rendered useless within 50 years by rising seas and increasingly powerful storms.

FLOODS (RAIN)

- **Flood control.** Cities can adapt to flooding from excessive rainfall by (1) constructing dams, (2) making flood-resistant buildings, and (3) diverting floodwaters. Dams can be temporary piles of sandbags, waterfront walls, or large engineering constructs straddling major rivers. (It should be noted that dams turn rivers into lakes, disrupting riverine ecosystems, causing species extinctions, and fostering the growth of anaerobic bacteria that emit methane.) Elevating roads and buildings allows floodwater to flow under structures and not through them. Semipermeable pavement, rainwater storage in low-lying areas (sometimes doubling as sports fields), underground floodwater storage tanks, rain barrels, drainage canals, large stormwater pipes (with separate systems for sewage), and extensive rain gardens to soak up excess water are some common methods for diverting floodwaters away from city streets. (They may not work for major floods.) Some cities are buying out homeowners in flood-prone areas. And occasionally small towns physically move to higher ground. This is what the village of Rhineland, Missouri (population 150), did after the devastating Missouri River flood of 1993.

- **Floating houses.** The Dutch are at the forefront of floating house design. Some of the homes they've built are basically fancy houseboats on the canals of Amsterdam. Others are attached to pilings that allow a home to rise with rising waters.
- **Better city planning.** In August 2017 Hurricane Harvey paused over Houston, Texas, dumping more than 4 feet (1.2 m) of rain. Tens of thousands of homes were flooded because they were located on floodplains defined by historical flooding events. The moral? Cities like Houston need to pass and enforce zoning laws that anticipate the floods generated by more powerful storms and other extreme weather events of the future. The past can no longer serve as a guide for floodplain zoning maps.

INFRASTRUCTURE

- **Share the Road.** This is a United Nations Environment Programme (UNEP) endeavor that helps developing countries reduce dependence on cars by building safe infrastructure for cyclists and pedestrians. Cities in wealthy countries (more so in Europe than in the United States) are expanding cycling trails, bike sharing programs, and car-free zones. The benefits include less pollution, less traffic congestion, fewer emissions, lower nationwide fuel costs, and a healthier populace (UNEP, n.d.).
- **Strengthening of infrastructure.** Tomorrow's hotter climate (especially in temperate cities unaccustomed to the heat) will buckle highways, expand joints in bridges, warp railroad tracks, weaken steel structures, and overload electrical grids. According to a Colorado State University study, one-quarter of American steel bridges will fail within 21 years because of climate change (Grossman 2019). That number approaches 100% by the end of the century if nothing is done to mitigate climate change. As a nation, we must invest in strengthening infrastructure.

Health

Shiver in my bones just thinking about the weather.
—Natalie Merchant, "Like the Weather"

Weighing the Effects of Climate Change

Suppose you built a scale (like the scale of justice) to measure the good and evil consequences of climate change. For every climate event the scales would tip heavily toward the evil side, but not completely, for the emerging climate occasionally brings a smattering of benefits (along with a whole mountain of malevolence). Heavy rains, for example, might cause devastating floods in one place and save a farmer's parched fields somewhere else. Warmer temperatures mean heat waves in the Texas Panhandle but a longer growing season in Alberta, Canada. Rising seas will drown coastal cities but will also open new habitats for marine life.

"But wait," you protest, "what about drought?" True, it's hard to see much good coming from a drought . . . unless you're a student of history. Droughts cause lake and river levels to drop, revealing formerly submerged artifacts, villages, battlefields, warships, and Indigenous campsites, allowing historians and archaeologists to learn more about the past. In fact, the warming climate has been a godsend for archaeology. Discovered under melting Arctic ice and in melting permafrost are Stone Age tools, Viking weapons, human remains, animal fossils, and other secrets now coming to light thanks to climate change.

To be sure, the future climate will be a disaster for most societies. But it will offer some boons as well, for everything that is . . . except our health. When it comes to the effects of climate change on human health, there are no benefits. It's all bad.

Disease

Diseases along with heat waves, storms, and wildfires are shifting through time and space. Climate change is altering the geographic range, seasonal distribution, and abundance of disease vectors. We can expect tick-borne illnesses such as Lyme disease to spread northward (see "Nature on the Move"). West Nile virus and the Zika and Chikungunya viruses will show earlier seasonality and a northern and upslope spread associated with warming temperatures.

Mosquito-borne illnesses will increase following extreme precipitation events and also during periods of extreme drought. That's because during droughts river levels drop, turning free-flowing water into stagnant pools, perfect breeding grounds for mosquitoes. It should be noted that malaria and some other mosquito-borne diseases can infect both people and animals.

There are also concerns that microbes preserved for centuries in frozen ground are coming back to life. The causes of the 2016 anthrax outbreak in Siberian reindeer included melting permafrost, higher summer air temperatures, and the discontinued vaccination of reindeer against anthrax. Anthrax spread from one thawing reindeer carcass to infect about 100 people. One child and at least 2,300 reindeer died (see "Permafrost" in "Earth").

Scientists debate whether the resurrection of frozen pathogens (the so-called *zombie pathogens*) is myth or reality. Not all microorganisms can survive extreme cold, not all viruses can last for decades, and the infectious agents buried in the permafrost are largely unknown and unpredictable in their emergence. While there may be some concern over zombie pathogens, a more significant worry is conventional diseases like malaria that will spread with warming temperatures.

Heat

High temperatures increase the risk of illness and death, especially cardiovascular and respiratory complications afflicting older adults. Heat waves and extreme weather can disrupt continuity of care, such as access to therapy or medication, thereby exacerbating mental illness. We can expect a decrease in premature cold-weather-related deaths but an increase in premature hot-weather-related deaths over the next century. Cramps, heat exhaustion, and heatstroke will become familiar news items in the coming decades.

As with many of the expected climate-change-related health problems, the young, the elderly, the chronically ill, some communities of color, and the poor will be at greater risk, as will those who work outdoors. In El Salvador an astounding 20% of the population (mostly men) have chronic kidney disease due to dehydration from working outside in temperatures much higher than in the recent past (Wallace-Wells 2020).

On the growing list of health issues spurred by a warming climate, we must now include sleep deprivation. New studies have found that the quality of our sleep diminishes in proportion to rising temperatures, in particular when air-conditioning is unavailable or not working. Without proper sleep, people can become depressed, angry, moody, and exercise poor judgment. For some reason, higher temperatures affect the sleep patterns of women more so than those of men (Weiss 2022).

Heat can make us edgy, grouchy, and angry. This we know instinctively. Think about the stress of getting caught in a muggy-summer-day traffic jam, cars crawling bug-like along congested roads, irate drivers muttering insinuations about another driver's unsavory ancestry. Road rage is already an issue today. How will that rage express itself in a more crowded and much hotter future?

Criminologists have long noted a link between high temperatures and aggression in humans, which can lead to a rise in local crime rates. Now there are studies attempting to quantify the impact that climate change will have on crime, in particular on violent crime. One such study, published in the journal *Environmental Research Letters* in 2020, suggests the United States could see 2 million to 3 million more violent crimes between now and the end of the century than there would be in a non-warming world (Harp and Karnauskas 2020). It is theorized that warmer winters will create more opportunities for interactions between people, thus putting more people at risk of being crime victims. Another theory is that people are more likely to read personal interactions as aggressive when experiencing higher temperatures. Of course, both theories could be true. Either way, the future is likely to be a violent one.

Humans aren't the only ones who get testy with high temperatures. A new study by Harvard Medical School found that dogs bite more often when the days are hotter and more polluted (Dey, Zanobetti, and Linnman 2023). The link between aggression in humans and heat is apparently the same with dogs. If the data holds true, what does this mean for other species? Will bear attacks be more frequent during heat waves? Should we be more cautious while hiking on hot days?

Food

Climate change will have a profound impact on the food we eat, not only in the distribution and availability of certain crops but also with food safety and nutrition. Some of this is discussed elsewhere in this book, but the issues are important enough to bear repeating.

Warmer temperatures will lead to food shortages, famine, rapid food spoilage, more mold, and more insect infestations. Foodborne bacteria like salmonella will flourish in tomorrow's warmer and wetter climate. Higher CO_2 concentrations will reduce the nutritional levels of fruits and vegetables. The food industry will be affected by crop failures, declining yields of seafood, and disruptions in food supply chains. Scientists have only recently begun to unravel the intertwined connections between a warming climate and food safety and nutrition (see figure 80).

Don't Drink the Water

Waterborne diseases are expected to increase due to climate change and the stress on aging water treatment infrastructure. Higher water temperatures are lengthening the season for toxic algal blooms in fresh water and in the sea, increasing the risk of exposure to algal toxins linked to a variety of illnesses. This is an issue for people. It is also an issue for wildlife (see "Water Supply" in "Water").

Heavy rain, storm surges, and salt water intrusion from rising seas will threaten low-lying wastewater, stormwater, and drinking water infrastructure, most of which were not designed to face the extreme weather of the future. This will lead to increased risk of exposure to water-related pathogens, chemicals, and algal toxins in recreational and shellfish-harvesting waters as well as in drinking water. Flooding can lead to greater runoff of pesticides and fertilizers, contaminating irrigation water, produce, rivers, and oceans. Often sewage is a component of floodwaters. Add sea level rise to the mix, and we also can expect contaminated drinking water from saltwater intrusion into aquifers and groundwater.

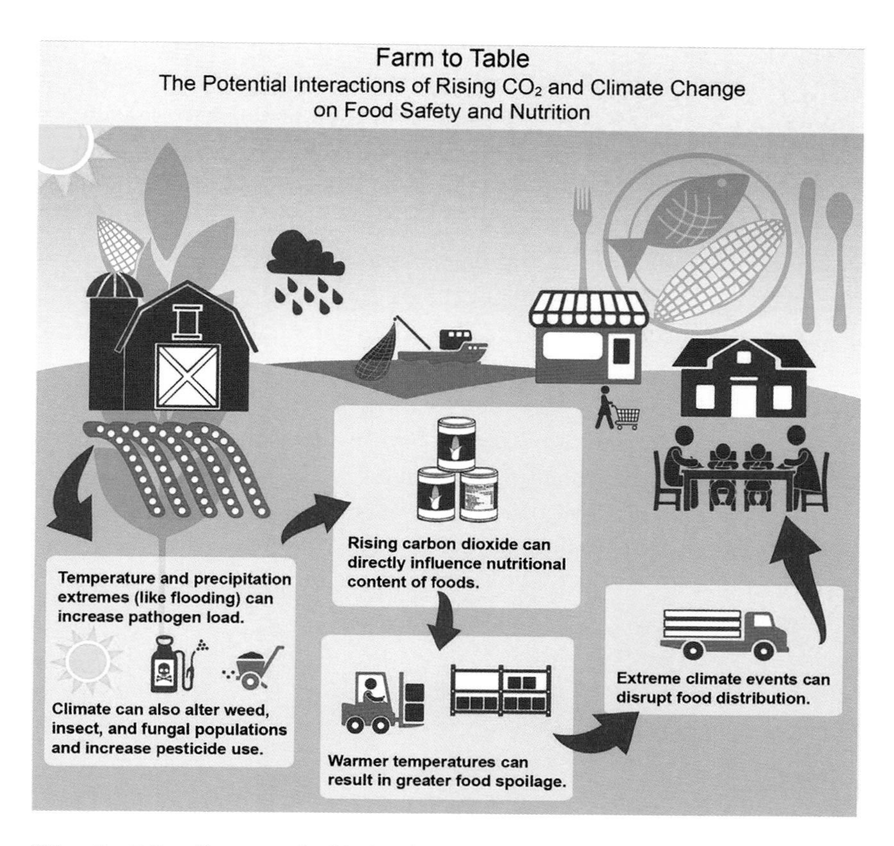

Figure 80. Follow the arrows in this drawing to trace the connectedness of our foods with climate change. Image from Climate.gov, US Climate Resilience Toolkit, Global Change Information System.

Don't Breathe the Air

You are what you breathe. And climate change will adversely affect what goes into your lungs. New wind and weather patterns impact the levels and locations of outdoor air pollutants such as ground-level ozone and fine particulate matter. We can expect more wildfires that release fine particulate matter and ozone precursors, resulting in increased premature deaths. Hotter temperatures, increased carbon dioxide (CO_2), and higher precipitation will generate more plant growth and therefore more pollen, fungal spores, and other allergens. Higher pollen concentrations and longer pollen seasons will exacerbate allergies and asthma, especially among the elderly. Deteriorating outdoor air quality means the same for indoor air. For some of us, allergy season is about to get a whole lot worse.

Numerous recent studies have found links between high concentrations of CO_2 and air pollution and cognitive functioning (see "Bad Air" in "Air").

This problem has been recognized in the past in lowered cognition from higher CO_2 levels for people wearing motorcycle helmets with face masks, or for NASA's astronauts in the International Space Station. But the connection to global climate change is only recently being emphasized. A 2018 University College of London study found that higher concentrations of CO_2 affect cognitive functioning and posited that, given the likelihood of increased CO_2 concentrations, direct impacts on human cognitive performance by the end of the 21st century may be unavoidable (Lowe, Huebner, and Oreszczyn 2018).

There are some who wish to downplay the potential impacts of climate change, and one thing they often cite is the increase in plant growth from higher levels of CO_2. However, increased CO_2 levels also lower the nutritional value of important food staples (see "Famine" in "Earth"). In addition, farmers will be challenged by changing weather patterns, including increased temperatures and weather extremes, droughts, and flooding. No matter how you play with the data, the climate scale always tilts toward the side of evil. Always.

Mental Health

Climate change will negatively impact mental health in numerous ways. Losing a family member from a flood or storm or losing a pet to a wildfire, for instance, can result in anxiety, depression, post-traumatic stress disorder, and suicide. The American Psychiatric Association warns that extreme weather events are also associated with increases in aggressive behavior and domestic violence (Morganstein 2019). The Association notes that particularly vulnerable populations include children, the elderly, the chronically ill, people with cognitive or mobility impairments, pregnant or postpartum women, people with mental illness, people of lower socioeconomic class, migrants, refugees, and the homeless. That's a lot of people.

Apparently some psychiatric medications interfere with a person's ability to regulate heat. Moreover, people with mental illness are more likely to live in poverty and to struggle with substance abuse, making it harder for them to cope with change or adversity. In addition, people being treated for mental health issues may be affected by climate disasters that, more often than not, disrupt their treatment schedule. Figure 81 illustrates the complexity of interactions between climate change and our physical and mental health.

Disaster Preparation for People with Chronic Illness

Surviving a climate-related disaster while dealing with chronic illness poses unique challenges and requires some extra preparation. A lot of the general

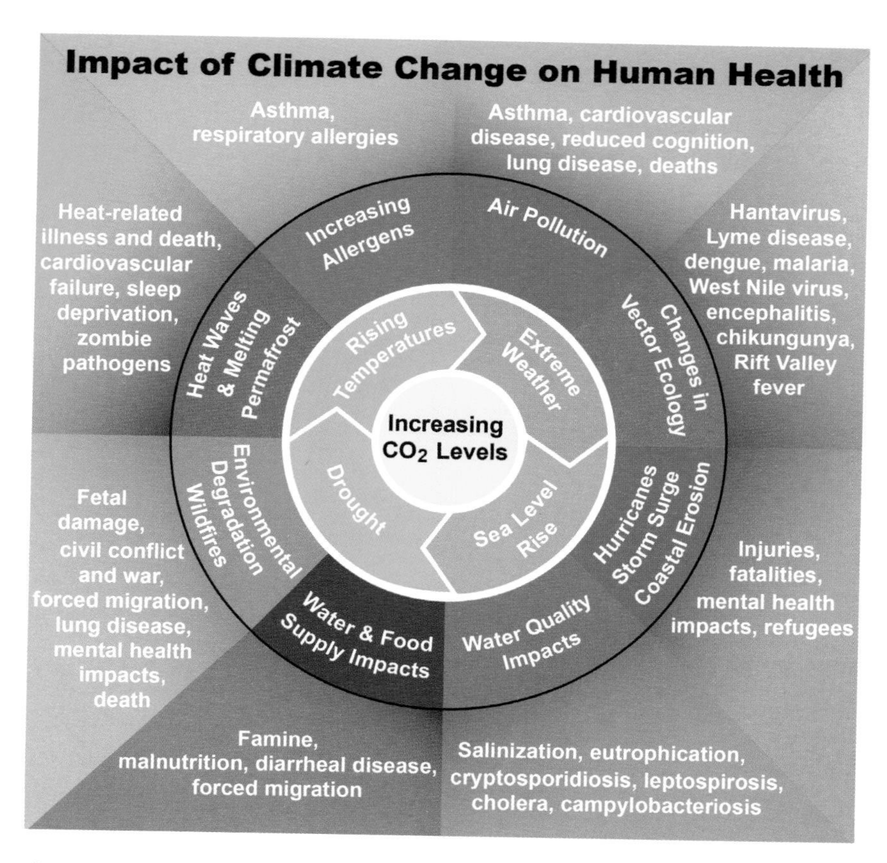

Figure 81. This chart illustrates some of the ways that climate change affects our physical and mental health. Image from Centers for Disease Control and Prevention, modified by Charles Pilkey.

advice still applies, but it is even more crucial that the chronically ill make plans prior to disasters. Isolation can be deadly in a disaster, so the first step would be to establish a self-help team to provide aid in the event of a disaster. These can be family members, friends, church members, coworkers, or neighbors. These should be people whom you can trust and who know about your needs. It is important to find several people who may assist you in your hour of need, in case one or more is unavailable. Make sure these people know how to administer the care you need.

The Centers for Disease Control and Prevention (CDC) recommend that one should try to have on hand a 2-week supply of prescription medications. It is also crucial to protect important documents such as a list of your prescriptions and providers, as well as contact numbers for friends or family. This list should be in your go bag (see "Bug-Out Bags") and recorded digitally on a phone or online. You want to gather enough food, water, and medical supplies to last at least 2 weeks.

You should have on hand medical devices that monitor health, such as a blood pressure cuff, oximeter, thermometer, and so on. Wear a medical alert bracelet to help first responders understand the chronic conditions you have and what type of care you may need. Be prepared for power outages, with flashlights, extra batteries, and a portable USB charger for your cell phone. If you use an electric scooter or wheelchair, have a backup power source on hand, and keep a manual wheelchair in case of power outage.

Before disaster looms, make evacuation plans. If you need assistance to travel, make plans well in advance with family and friends. As noted above, isolation can be deadly in a disaster, and it is important to make crucial community connections before disaster hits. For instance, in extreme heat events, power may fail, and you will lose your air-conditioning, or you may not have air-conditioning in the first place. If you live alone or need assistance, it is crucial to have plans for someone to assist you in getting to someplace cool, such as cooling centers or shelters the community sets up. Find out what places your community has designated as shelters for use in emergencies.

The Americans with Disabilities Act defines a service pet as an animal that has been trained to do work or to perform tasks for persons with a disability. However, there may also be state laws that apply as well. Familiarize yourself with state laws to make sure your pet qualifies as a service animal so that your pet may accompany you in the event of evacuation. Pets, other than service animals, may not be allowed in shelters.

Diabetics should keep on hand 2 weeks' worth of supplies, including extra batteries for the glucometer, syringes, strips, and alcohol wipes. An empty plastic bottle can serve as a sharps container for needles and lancets. An additional challenge for diabetics and other patients is that some medication requires refrigeration. Consider purchasing a cooling wallet to keep insulin and other medications cool. The manufacturers of the FRIO insulin cooling case claim it can keep temperature-sensitive medications cool and safe, within safe temperatures of 64.4°F – 78.8°F (18°C – 26°C), for a minimum of 45 hours, even in a constant environmental temperature of 100°F (37.8°C). The cooler works by being immersed in cold water, which activates a cooling gel that can be reactivated again for either continuous or intermittent use.

What to Do

- **Heat.** Stay hydrated; wear lightweight, loose clothing; and use an air conditioner or fan. If you have no air conditioner, go to a library, shopping mall, or any air-conditioned public space for temporary respite (see also "Heat" in "Air"). Walk your dog in the cool of the morning and evening, lest Fido gets snappy.

- **CO_2 and cognition.** A 2016 Harvard study found that the average CO_2 level inside American buildings and cars impairs decision-making skills. The study found that cognitive scores were higher in "green" buildings as compared to conventional buildings (Allen et al. 2016). We need to design the buildings of tomorrow to mitigate the impact of CO_2 and other pollutants. Whenever possible, open windows and let some fresh oxygen waft through your house.
- **Road rage.** Put a sign on your vehicle's dashboard: "Stay calm and listen to the music."
- **Polluted air.** See figure 32 (the DIY air filter in "Bad Air") to learn how to make a cheap air filter to improve the air quality inside your home.
- **Mental health.** Ours is an age smothered in the gloom of imminent climate disasters, our anxiety deepened by a barrage of disaster films, foreboding news reports, and dystopian, end-of-the-world cli-fi novels. This must be hard on those susceptible to depression. It's hard on anyone who is paying attention to the science of climate change. But the Earth still spins on its way around the sun. Flowers still bloom. There is still beauty in the world. And the best reaction to existential angst is action. As a nation we must do what we can to help our most vulnerable populations. We should help ourselves as well. A quiet walk immersed in Nature can do wonders for one's mental and physical well-being. Just remember to do it early in the morning . . . before the heat comes!
- **Disease.** More research is needed to predict how disease might spread in a warming world and to anticipate tomorrow's new health issues.

Nature on the Move

I took a walk in the woods and came out taller than the trees.
—Henry David Thoreau

The Great Migration

The nonhuman citizens of Earth (the ones we call Nature) have a key advantage over humans when it comes to adjusting to climate change. They don't require political consensus, research grants, or clever talk show hosts holding forth about the end of the world. Plants, animals, fungi, and microbes adapt to environmental change . . . well . . . naturally. They are doing so now, just as they have always done, by moving to more suitable locations, modifying their behavior, or evolving into new species (speciation), strategies governed by one simple, brutal maxim . . . adapt or die. We might do well to remember that maxim.

Species come and species go, and sometimes species go extinct in response to global change. It has always been thus. Plants, animals, and even entire ecosystems have migrated toward the equator or the poles, driven by the dictates of a fluctuating climate. If the climate changes gradually, then species can adjust accordingly. But if the climate cools too fast or heats up too quickly (as is happening now), then species are at risk of extinction (see "The Lessons of Geologic Time" in "Earth").

The closest analogue to today's climate was the mid-Pliocene, roughly 3 million years ago, when atmospheric carbon dioxide (CO_2) levels were about the same as today (around 400 parts per million). Temperatures were on average 3.6°F to 5.4°F (2°C to 3°C) warmer than today—more or less where scientists expect temperatures to be by 2100. So warm was the Pliocene that the poles were ice-free and tropical animals like crocodiles, hippos, and tapirs had moved north into the Arctic Circle. Palm trees grew in both the Arctic and Antarctic landmasses. The temperate deciduous forests now blanketing the Southeastern United States had relocated to Greenland and northern Canada.

Around 2.6 million years ago the world became cooler, triggering a succession of ice ages during the epoch we call the Pleistocene. Expanding ice sheets in North America and Eurasia chased life south to warmer climes. Mammoths made it as far as southern Florida, while dire wolves, reindeer, musk oxen, and camels roamed the American Midwest. During the Pleistocene's warmer, short-lived interglacial periods life moved north, only to move south again when temperatures cooled. Scientists believe the Earth is currently in an interglacial period, set to cool into another ice age. But the rise of industrial civilization disrupted the normal ebb and flow of the geologic seasons . . . and that has made all the difference.

Now species are on the move again, responding to anthropogenic climate warming, following ancient migration patterns as old as life itself, moving north in the Northern Hemisphere, south in the Southern Hemisphere, and upslope wherever possible. Gentoo penguins have expanded their range south to new parts of Antarctica. Fishermen in Alaska are catching warm-water species they've never seen before. The Arctic tree line in Norway is advancing poleward as much as 160 feet (50 m) per year. And trees are moving upslope in the Canadian and American Rockies, the Peruvian Andes, the French Pyrenees, and the Altai Mountains of Central Asia.

Trees like mangroves can migrate with relative ease. Those in Florida are doing so now, casting seedpods on ocean currents to colonize shorelines further north, made hospitable by a warming climate. But bristlecone pines (*Pinus longaeva*), perhaps the oldest living trees on the planet (some are 5,000 years old), are having a hard time ascending the eastern slopes of the Sierra Madre Mountains. Limber pines (*Pinus flexilis*) are marching upslope much faster (faster by tree standards), blocking the seed dispersal of their older cousins and possibly dooming them to extinction. When it comes to surviving climate change, fortune favors the swift.

The migration of Nature to more habitable areas is a good thing, boosting the chances of survival for some species that might otherwise succumb to the effects of heat and drought. But species trying to relocate to cooler habitats will be challenged by natural barriers (deserts, oceans, mountain ranges, and the like). They will also be tested by barriers put up by people. Animals bound for the promised land of soothing waters and milder temperatures must first cross long miles of urban sprawl and farmland. One can imagine the roadkill as increasing numbers of mammals, snails, turtles, snakes, and crawling insects negotiate treacherous highways and parking lots. Or the pileup of animal migrants at the US-Mexico border should a Brownsville-to-Tijuana border fence be constructed (see figure 82 in "Walled Out").

There is another kind of "migration" worth noting. As temperatures continue to rise, farmers in temperate regions may be able to grow crops once restricted to warmer lands (unless, of course, climate change makes farming there untenable as well). We see this happening in England and Scandinavia, where new vineyards are taking root while the traditional wine regions of France have suffered shrinking harvests from heat waves and extreme weather. In northern Canada, farmers are now growing corn and soybeans where formerly only cold-weather crops like wheat were possible. Coffee is being commercially harvested in California as traditional coffee producers in Brazil and Indonesia see declining yields due to higher temperatures. The climate shifts and crop belts do the same, making one farmer's loss another farmer's gain.

Invaders from the Tropics

One of the unwanted consequences of climate change is the migration of invasive species to new habitats, harming ecosystems by taking over the space native species would normally occupy and in some cases by transmitting disease. Disease-carrying insects like mosquitoes are flying out of the tropics into temperate regions, infecting people with malaria, Zika virus, dengue fever, yellow fever, West Nile virus, and other diseases. Lyme disease, already common in the United States and southern Canada, is expected to reach the Northwest Territories later this century as deer ticks advance north with warming temperatures.

New contagions may be coming out of the Arctic as well. In 2016 anthrax infected dozens of Siberians, spreading from a 75-year-old contaminated reindeer carcass that had thawed during a heat wave (see "Permafrost" in "Earth"). Should we expect similar outbreaks (of so-called zombie pathogens) to happen in the future as more disease-ridden animal (and human?) carcasses emerge from melting permafrost?

Invasive species arrive (1) by hitching rides on planes and cargo ships, (2) through intentional or inadvertent release by people, or (3) by migration into

Figure 82. The Quino checkerspot butterfly (*Euphydryas editha quino*) is one among thousands of species that will be threatened by walls built to stop the hundreds of millions of human refugees expected to cross national borders this century. The Quino checkerspot lives on the Mexico-California border. A border wall would isolate populations in the United States from those in Mexico, diminishing the species' chances for survival. Photo by Andrew Fisher / US Fish and Wildlife Service volunteer biologist.

Walled Out

Once numbering in the millions, the Quino checkerspot butterfly is now a federally listed endangered subspecies. Because it rarely flies higher than 15 feet (4.6 m), a 30-foot (9 m) border wall along the southern US border will hamper its movement back and forth between California and northern Mexico. Such a wall will also threaten jaguars, jaguarundis, ocelots, Mexican black bears, pygmy owls, Sonoran pronghorns, and a host of other species trying to escape the heat by moving north into the United States. As the climate heats up, hundreds of millions of human refugees from Africa, Latin America, and Asia will join the north-moving animal exodus. Governments worldwide will be pressured into constructing border walls. People of course can climb walls. Most animals can't.

new habitats (now facilitated by climate change). Chestnut blight, lionfish, fire ants, Asian carp, zebra mussels, lampreys, and Burmese pythons . . . such are the poster children of invasive species devastating North American ecosystems. The devastations will only get worse as species like the Burmese python methodically eat their way north (see figure 83).

The Insect Apocalypse

About 87% of plants rely on birds, mammals, and insects for seed dispersal. As seed-dispersing animals decline in numbers from local habitat loss or migration to cooler lands, plants will have fewer pollinators on hand. Espe-

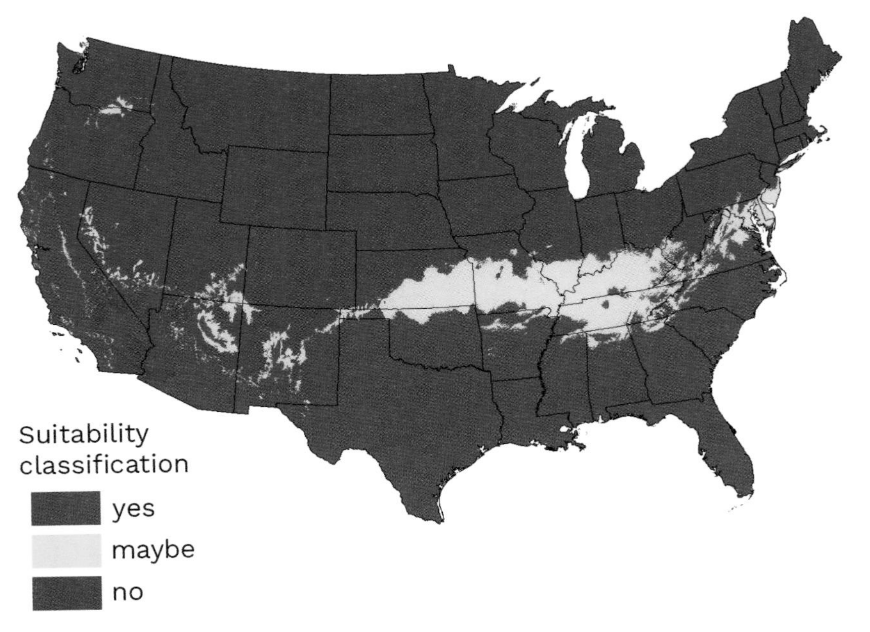

Figure 83. Map showing the possible distribution of the Burmese python by 2100, based solely on the likely temperatures at that time. The map assumes that the snake, a native of Asia but in the United States a resident of the Everglades, will migrate northward along with the increase in atmospheric temperatures. Photo from US Geological Survey.

cially problematic are those plants pollinated by insects. That's because insects are vanishing at an astonishing rate. In just 50 years the world's insect biomass has declined by as much as 75%, an extinction event that some are calling the "insect apocalypse."

Why insects are dying is a mystery, but it seems likely that habitat loss and pesticides are the main culprits. Light pollution is also a known cause of insect decline, but the causal connections are still being worked out by scientists. Because most of us live in cities, disconnected from the natural world, we are only vaguely aware of the critical role insects play as decomposers; as food for fish, reptiles, mammals, and birds; and as pollinators of our fruits and vegetables. We are even less cognizant of their dwindling numbers.

But there is some anecdotal, albeit unscientific, evidence that makes clear the magnitude of insect decline. Anyone old enough to remember driving in the 1960s and 1970s will doubtless recall having to make frequent stops to clean windshields, hopelessly soiled with the splattered corpses of beetles, butterflies, and other flying insects. Today we can drive all day with windshields largely unsullied and bug free. The *windshield effect* (or *windscreen effect*), as it is sometimes called, is seen by biologists as a prelude to a looming disaster for insect-pollinated plants and for those organisms that feed on insects, more intimations of a biosphere starting to unravel.

Shapeshifters and Shifting Behavior

Earth's climate entered a warming phase near the end of the Pleistocene until finally stabilizing around 10,000 years ago, making possible the invention of agriculture and the subsequent rise of civilization. Now that 10,000-year period of relative stability is gone. The seasonal rhythms we grew accustomed to are fast disappearing, creating massive ecological mismatches. Flowers are blooming earlier in the spring, before their pollinators arrive. Bears are waking sooner from hibernation, when their usual food sources are not yet available. Birds are going hungry because the insects (the ones that haven't already died out) they normally eat are emerging earlier and are gone before the birds' chicks can hatch. Timing is everything in Nature, and the timing today is way off.

Fortunately, animals have varying degrees of behavioral plasticity, allowing some species to cope with a changing climate. A few pika populations, contrary to the general upslope movement of most animals, are descending from the fast-warming peaks of their normal habitat to live in the cool forested valleys, changing their dietary preferences along the way from grass to moss (see figure 84). They have even learned to extend all four legs horizontally to give off excess summer heat.

Some Arctic birds like dovekies have also devised new feeding strategies as their old food sources at the edge of ice floes have moved elsewhere with retreating ice packs. And then there are the fish off southeastern Alaska and New England, which are moving into deeper water to avoid heat stress.

Other species are adapting to climate change by physically altering their bodies through genetic changes. In a word . . . they are evolving. Something scientists used to think required thousands or even millions of years is unfolding before our eyes. Whales, fish, birds, frogs . . . animals everywhere are evolving, becoming smaller in response to hotter temperatures (smaller bodies dissipate heat faster). Appalachian salamanders are 8% smaller than before. Fully grown Humboldt squid are now 50% smaller. Even birds in the Amazon are getting smaller. At the same time, birds are evolving longer beaks and longer feathers and wings to help rid their bodies of excess heat. Insects' legs are getting longer for the same reason. On the Turks and Caicos Islands, anoles (gecko-like lizards) are evolving stronger toe pads and shorter legs to better anchor themselves to trees as increasingly powerful hurricanes hammer the Caribbean. Those with longer, weaker legs get blown away, effectively eliminating their DNA from the local gene pool.

Figure 84. Pikas are an example of an animal able to change feeding habits to accommodate a changing climate. Normally they live high on rocky peaks, but some now are feeding in the dark forests on the valley floors below the peaks. Photo from the US Geological Survey.

One can quibble over whether these genetic changes mark the advent of new species, but it is incontrovertible that we are witnessing evolutionary processes at work vis-à-vis climate change. Natural selection on a warming planet is a hard taskmaster, favoring the smaller, the tougher, the swifter, and those able to change behavior in a way that enhances survival in a rapidly changing world.

What to Do

- **Learn from Nature.** We need to follow Nature's lead by migrating to cooler places and by changing our behavior (driving less, eating less meat, etc.). We are not likely to evolve into a new species (at least not anytime soon), but we can "evolve" new technologies to cope with a warming world.
- **Create Nature corridors.** Research has demonstrated the effectiveness of establishing corridors through which animals can move and plants can disperse their seeds. These include wildlife crossings under or over highways, greenways, and interconnected

wildlife refuges. Border walls should be designed to allow Nature free passage (or not built at all).

- **Prepare for pandemics.** We need to prepare for the inevitable spread of new climate-related diseases by (1) funding epidemiological research, (2) expanding our health infrastructure to monitor the spread of disease, (3) developing new vaccines for diseases like West Nile fever, and (4) developing a warning system, perhaps along the lines of the emergency weather alerts on cable TV and radio.
- **Assist Nature.** To avoid ecological collapse, we should do the following: (1) physically move endangered plants and animals to safer locales, (2) genetically modify organisms (assisted evolution) to make them more resilient to climate change (Yes, we recognize the inherent moral issues and potential dangers, but what is the alternative?), (3) restore native ecosystems (aka rewilding), (4) eliminate the most damaging invasive species, and (5) stop deforestation everywhere, including the razing of urban forests for the construction of subdivisions, factories, apartments, and shopping malls. We should reconsider the custom of cutting live trees for Christmas. The more trees that grow to maturity, the more carbon is pulled from the atmosphere.
- **Develop new commercial foods.** This is a global issue. Farmers should anticipate changing climate patterns and select the kinds of crops they should grow accordingly. Fishermen should expect diminishing catches, new species appearing in traditional fishing grounds, and the possibility that they may need to seek other employment.
- **Preserve insects.** Reduce the use of insecticides, especially in parks and in people's yards. Stop using bug zappers. They kill everything, including insects that don't bite. We need to teach our kids not to fear insects but to respect them for their various ecological roles. Insects are awesome!
- **Consider de-extinction.** In 2003 scientists for the first time revived an extinct species. The Pyrenean ibex (actually a subspecies of the Iberian ibex) went extinct in 2000. Through a complex cloning process, a living specimen was born only to die a few minutes later from a lung defect. Scientists around the world are trying to bring back extinct species through cloning and genome editing. Species slated for resurrection include passenger pigeons, Tasmanian tigers, aurochs, dodos, and mammoths. Aside from the ethical issues (Did anyone ask a mammoth if it wanted to be reborn?), de-extinction, or resurrection biology as it's sometimes called, will probably happen

within a decade or so. As with any new technology, no one has fully considered the consequences: Can species brought back from the dead survive in a world different from the one they evolved in? What about resurrected species like whales or apes that possess what can only be called culture? Can you resurrect a species' language as well? Entire books have been written on the bioethics of de-extinction. More will be written in the decades to come. Still . . . wouldn't it be nice to see a living Carolina parakeet? Or a mammoth?

The Biosphere

We need ants to survive, but they don't need us at all.—E. O. Wilson

The Evolution of the Biosphere

The biosphere . . . the zone of life and sum of all ecosystems. The womb of the Earth. A sphere with a surface 12.5 miles (20 km) or more thick teeming with life from deep ocean trenches to the upper atmosphere, even into solid rock miles underground. A network of interacting, coevolving systems (decomposers, symbionts, producers, consumers, etc.) that in concert with geochemical cycles has maintained, sustained, and regulated a habitable world for billions of years (see figure 85).

Life on Earth began not so long after the oceans formed, perhaps 4 billion years ago through a process called *abiogenesis*—life rising from nonliving matter. The first organisms were prokaryotes that resembled simple anaerobic bacteria. Around 2.5 billion years ago, some prokaryotes called *cyanobacteria* developed the ability to photosynthesize, giving off enough oxygen to create our oxygen-rich atmosphere, an episode termed the Great Oxygenation Event. A billion years later, multicellular life-forms appeared, evolving

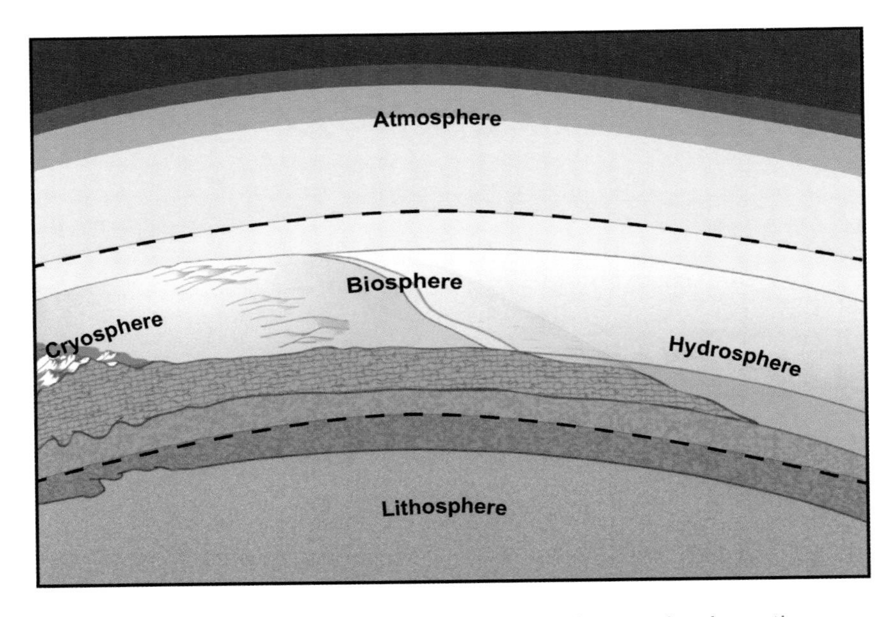

Figure 85. Scientists conceptualize the Earth as a series of concentric spheres: the lithosphere (crust and upper mantle), the hydrosphere (water), the cryosphere (ice), the atmosphere, and so on. The biosphere (the area between the dashed lines) consists of overlapping biomes (rainforests, deserts, coral reefs, and the like). Its borders are ill defined. Scientists have not yet determined how high in the atmosphere or how deep in the crust life exists. Drawing by Charles Pilkey.

into various groups that became the common ancestors of all plants, fungi, and animals living today (National Geographic Society, n.d.).

The biosphere is thus the product of gradual planetary evolution, characterized by an increase in the numbers, interdependence, and diversity of life over stretches of geologic time. It has proven to be remarkably robust, having survived multiple mass extinction events. Doubtless it will rebound once more from the current episode of human-induced extinctions. But this doesn't let humanity off the hook.

Our planet's biosphere today little resembles its antecedents. We have created entirely new ecosystems (cities, farms, and highways), eliminating along the way 83% of all wild mammals and 50% of all Earth's plants. The current extinction rate is 100 to 1,000 times the normal background extinction rate. We are eliminating species through habitat loss, the introduction of invasive species, overfishing, overhunting, and the spreading of industrial pollutants (more or less in that order of influence). Climate change (including ocean acidification), which has not been a recent cause of biodiversity loss, will soon become a major contributor of species extinctions, especially if Earth's coral reefs disappear as expected sometime this century.

In addition to species loss, the beauty, recreational pleasure, and spiritual wonder that many seek in Nature are drowning in a flood of development. Even the wild, open spaces of the far north and those in Patagonia will likely degrade as hundreds of millions of climate refugees relocate to cooler climes (Carrington 2018).

Some extreme proponents of capitalism argue that the Earth and everything in it is ours to commodify as we please. After all, we are the most intelligent life on the planet, the only species to craft cars, computers, and moon landings. Species extinction in this view is an unfortunate but inevitable consequence of progress. But does technological might make right? Do not plants and animals have an intrinsic right to exist?

When a Tree Is a Person

Over time our concept of justice has expanded in ever-widening circles of ethical concerns, from the entitlements of the wealthy to the rights of women, Indigenous people, and other minorities. The next logical stage in this development is the rights of Nature, or what might be called *biospheric justice*. This is no theoretical supposition. The movement has already begun.

It began in 2008 when Ecuador became the first country in the world to codify in its constitution the rights of Nature (Surma 2021). This allowed citizen groups to legally petition on behalf of Pachamama (Mother Earth). Other countries followed Ecuador's lead. Bolivia soon passed similar laws. In 2017 New Zealand declared the Whanganui River to be a person. India did the same for the Ganges. Pakistan, Colombia, Argentina, Peru, and several towns in the United States and Canada have either passed laws granting personhood to Nature or invoked the concept of the rights of Nature to protect local ecosystems. More recently, in September 2022, Spain granted legal status (personhood) to Mar Menor, a large saltwater lagoon on the country's Mediterranean coast (J. Wilson 2022). The movement continues.

Granting an ecosystem the same legal rights as a person requires a radical paradigm shift in our view of the human/Nature relationship. National governments, corporations, and individual landowners will resist such a transition, protesting loudly about job loss and infringement of personal liberty. But legal protection of Nature has given us the Grand Canyon and Yellowstone and may yet be the best way to preserve what wilderness still remains. To rephrase Scripture, what does it profit a corporation to gain the whole world but lose its soul?

The legal justification for the rights of Nature is twofold. The first is argument by analogy: we grant human rights to children even though they can't defend themselves in court. The same should be true for plants and animals, for the ecosystems they inhabit and by extension for the entire biosphere.

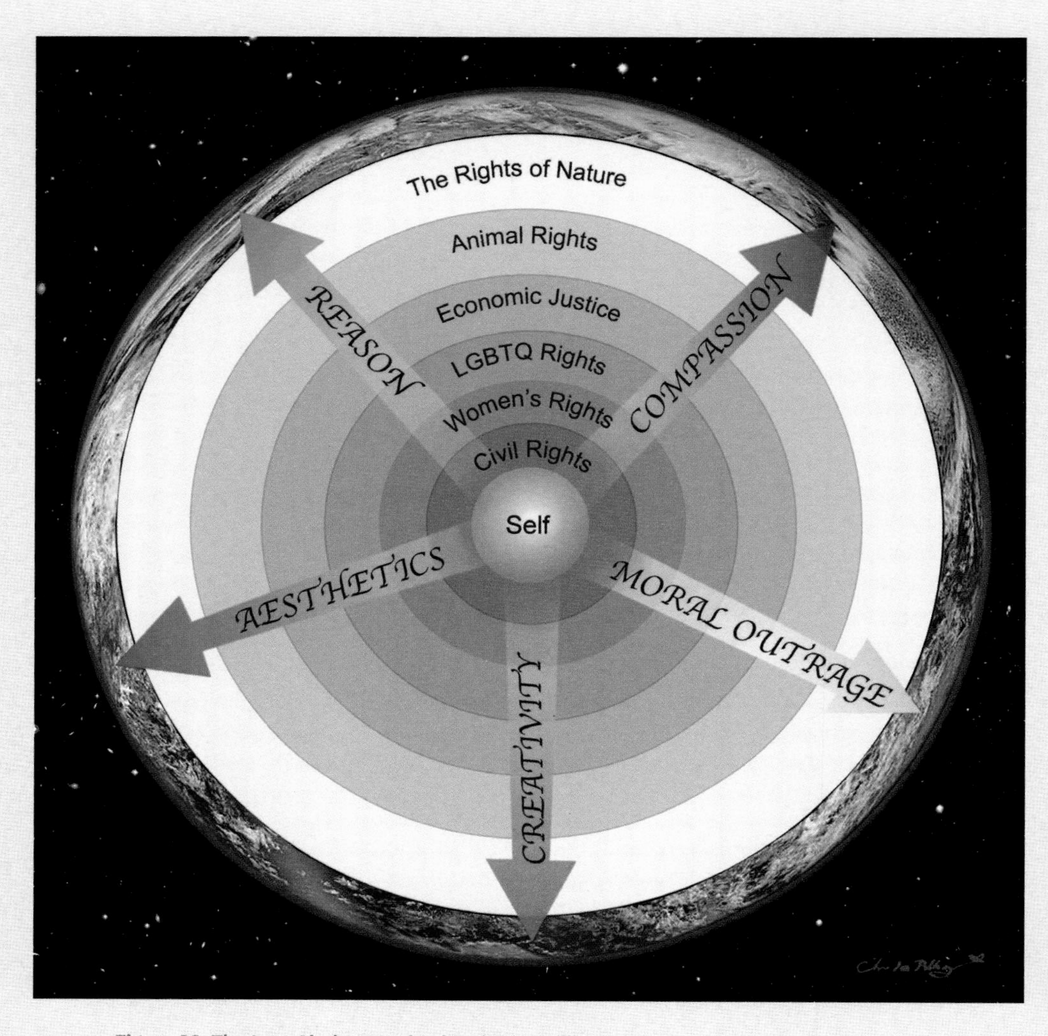

Figure 86. The Last Circle. Drawing by Charles Pilkey.

The second justification is one of practicality. Human health (some would say survival) is linked inextricably to the health of the biosphere. The two are one and the same. Civilization is as much a part of the biosphere as a community of lions in the Kalahari. It is only our language that discriminates, generating the illusion that people and Nature are separate entities.

In 2021 Ecuador's law was put to the test when the town of Santa Ana de Cotacachi filed a lawsuit against the state mining company ENAMI and its Canadian partner Cornerstone Capital Resources to prevent gold mining in the protected Los Cedros cloud forest of northwestern Ecuador. Amazingly,

The Last Circle

Social transformation generally comes from the bottom up, from citizens driven by a sense of moral outrage over injustice of one kind or another. Other transformative factors come into play as well. Our innate compassion, creativity, and ability to reason help focus energies toward resolving environmental and social ills. And then there's the important role of aesthetics, informing us of the beauty of untrammeled wilderness and inspiring with visions of utopia.

The diagram in figure 86 illustrates the evolution of our legal framework from the rights of humans to the rights of Nature, shown as circles expanding over time to include larger segments of society and eventually of the biosphere. Economic justice refers to capitalism, which in its current phase fosters disparity of wealth as well as the commodification and ruin of Nature.

Studies have shown a link between income inequality on one side and species extinction, higher per person carbon emissions, and higher rates of deforestation on the other. No one disputes the benefits of capitalism, but economic injustice and the degradation of the biosphere will persist until we fashion an economic system that equitably shares wealth while sustaining the well-being of plants and animals (Boyce 2021).

A healthy ecosystem maintains soil, recycles nutrients, and provides us with food, medicine, clean water, purified air, and raw materials like wood for our homes. It is the foundation on which civilization is built. Moreover, the greater the diversity and health of a given ecosystem, the more quickly it can recover from wildfires, droughts, and other ravages of climate change (Dunn 2021).

Taming the excesses of capitalism and developing green technologies will go a long way toward the restoration of damaged ecosystems. But it will not be enough. We must also move our ethical concerns outward toward the last circle . . . the rights of Nature. And the sooner the better.

the law prevailed, and the companies were prohibited from even exploring the region's mineral wealth.

Judge Agustín Grijalva Jiménez wrote the majority opinion, including the following: "The very existence of humanity is inevitably tied to that of nature. . . . Therefore, the rights of nature include necessarily the right of humanity to its existence as a species. It is not a rhetorical lyricism, but a transcendent confirmation and commitment that, according to the preamble to the Constitution, requires 'a new form of citizen coexistence, in diversity and harmony with nature'" (quoted in Surma 2021).

The Great Silence

Many have puzzled over the riddle of the Fermi paradox: If the universe is so vast and therefore packed with intelligent life, why has no one contacted us? Why no visitations from the deep void? Why the great silence?

Some say the immense distance between stars precludes such contacts. Others contend the technology of space travel is too complex for most life to develop or that humanity is too technologically backward, primitive, and aggressive to attract the attention of any star-roving species.

But there is another thread to the discussion, a dark possibility sometimes mentioned by science fiction writers. It goes like this: Intelligent beings evolved on countless worlds and developed advanced technological civilizations, inventing along the way complex mathematics, philosophies, and arts. Then they destroyed themselves in war and/or ecological ruin, collapsing into the ignominy of extinction. Given our own history of ancient civilizations (the Romans, the Mayans, the Khmer, etc.) that self-destructed from environmental degradation, overpopulation, climate change, and war, the theory that alien civilizations might do likewise seems a reasonable proposition (Diamond 2005).

History may not repeat itself, but it does rhyme. And it has rhythm. The first wave of 15th-century Europeans sailing to the New World was financed by government, the next wave by wealthy merchants, both groups pushed by militant nationalism and mercantile ambitions as much as by curiosity to see what lay beyond the next horizon. The same pattern unfolds today with the exploration of space. Space capsules initially launched with government funding are now being sponsored by corporate patrons, motivated partly by scientific curiosity but also by the lure of extreme wealth. When Spanish ships returned with New World gold, the Crown became rich beyond a beggar's dreams. When the first asteroid is towed to Earth, someone will become the world's first trillionaire.

But there is one essential difference between the previous age of discovery and today's move into space. When Columbus arrived in the Americas, there were air, water, food, and warmth for sustenance. The first Mars explorers will step onto a dead world sustained by whatever technology they bring.

Some technocrats proclaim that humanity's near-term survival requires moving off world to colonize Mars. But it took a billion and a half years for a biosphere to appear on Earth. It will take centuries to terraform Mars to the point where people can live there without a constant flow of terrestrial supplies. Physics is an implacable taskmaster, and no amount of wishful thinking can change its laws.

And so we arrive at our own moment of clarity. There is no Martian sanctuary, no refuge in space... nor even a safe anchorage on Earth to ride out the climate storm indefinitely. The strategies presented in this book are at best temporary expedients, ways to adapt to the consequences of a warming planet, to reduce suffering and hopefully buy enough time for humanity to deal with the root causes of climate change. Adaptation without mitigation leads to the disastrous endgame of biospheric collapse. Not an option.

This is not to be construed as an argument against space exploration. Humanity should explore and eventually colonize other worlds. But Earth comes first. We must invest sufficient resources to restore ecosystems, preserve biodiversity, and eliminate carbon emissions (including those from space launches). We had better start acting now like the stewards of Pachamama we claim ourselves to be. If we fail to keep our home world intact, then we fail as a species and the human experiment comes to an inglorious end. Game over.

And the rest is silence.

What We Did

A gathering of tourists at an overlook... a restored Alberta prairie... scattered islands of oaks, maples, and redbuds in a sea of grass taller than a tall person can reach... stretching in all directions.

"Hard to believe all this used to be farms and cattle ranches."

"Well, it helped," said a woman, turning to face the man who had spoken, "that we cut our population by almost a third, right?"

"Fewer people... more Nature," agreed the man. "But I think switching to a vegetarian diet really made the difference. It curbed methane emissions. Ranches returned to forest and prairie. You know... natural carbon sinks."

"Is it true," the woman hesitated, "... people back then ate animals?"

"They did. They even used the body parts of animals like tigers to make fake medicines."

"Disgusting! I'm so glad we live in a more enlightened age!"

"Everything changed," said the man, "when nations legislated the rights of Nature. Gave Nature a fighting chance. That's why we have all this." The man gestured across the grassy expanse toward a herd of buffalo grazing in the distance.

"By the way, are those your kids playing under the trees?"

"They are," smiled the woman. "Two of the three are adopted, of course. We're doing our part to control population. Aren't they beautiful?"

"They are. A beautiful rainbow mix."

"Their father's family immigrated from Nigeria. Refugees like so many others of the African diaspora... fleeing the heat."

"You know," said the man, "I may not look it. But I'll be 130 this fall. I remember well the rampant racism and social injustice of the 21st century. Mixed families like yours would have been, shall we say . . . outside the norm?"

"A hundred and thirty . . . really? You don't look a day over 60."

"The magic of longevity drugs . . . and a healthy lifestyle. But I was actually born before the Great Collapse."

"No way!" said the woman. "You lived through the Collapse?"

"Hungry times," said the man, looking vaguely into the distance. "Crop failures, food riots, wildfires, storms . . . wars. You don't want to know what we ate to survive!"

"That's amazing. You saw history as it happened."

"People were so materialistic back then," he said. "They could not live simply. Had to buy fancy cars, big houses stocked with all manner of useless goods. Pumped carbon into the air like there was no tomorrow."

"Back then," he continued, "people used to talk about escaping Nature. But it was really ourselves we had to escape from . . . our greed, our shortsightedness."

"Things are sure different now," said the woman. "We reinvented capitalism to help the poor and restore ecosystems."

"They call it 'regenerative capitalism,'" said the man. "But it should have happened sooner. We understood the links between income disparity and biodiversity loss for decades."

"What do you think transformed society?"

"Sea level rise," said the man, chuckling softly. "The sea rose up like a watery demon. Took out Miami, Shanghai, much of Florida . . . the Outer Banks. That was the wake-up call. The same politicians who ignored social and environmental issues for so long started pushing renewables . . . solar, wind . . . fusion and the like."

"They also taxed the superrich," said the woman.

"They did," said the man. "Finally."

"Do you think people will ever live in the south again?"

"I do," said the man. "Plants and animals are reversing course . . . starting to migrate toward the tropics. The climate appears to have stabilized. The sea is still rising but not so fast as before."

"You know, my husband," said the woman, "is the chief engineer for that." She pointed to a vertical structure standing on the horizon, a carbon capture plant, one of thousands of similar plants worldwide.

"Why don't you join us for lunch? I'm sure he'd love to hear your life stories."

"Thanks," said the man, "but I'm on my way into town to sign up."

"Sign up?"

"For the next Mars Shuttle," he said. "I'm a sculptor." He made a motion with his hands as if hammering a chisel.

"I figure my work here is finished," he continued, "now that the biosphere's on the mend. I'm going to be the first human to carve stone on Mars!"

"Well, good luck to you, then," said the woman.

"And to you as well," said the man. "Blessed be the Earth."

"Amen," said the woman.

THE HEART OF THE MATTER

Climate change is already causing floods, rising seas, wildfires, melting of permafrost, droughts, heat waves, increased storm intensity, ocean acidification, crop failures, growing numbers of refugees, famine, military conflicts, and widespread damage to ecosystems. While we wait for countries to reduce carbon emissions, we must adapt to an increasingly hostile climate.

Earth

- All five mass extinctions of the past 500 million years were caused by climate change. Scientists think anthropogenic climate warming if unchecked will precipitate a sixth mass extinction, killing off 75% or more of all Earth's species.
- The 2021–22 United Nations (UN) Climate Reports issued a "code red for humanity" concluding that (1) climate change is already happening, (2) it's caused by humanity's greenhouse gas emissions, (3) things will get worse for centuries to come even if we reduce our emissions now, and (4) the rate of climate change is accelerating (IPCC 2022, 2021a, 2021b).
- The regions most affected by climate change are among the poorest and most marginalized but are the least responsible for carbon emissions.
- A hotter climate will lead to crop failures, diminished fish stocks, increasing crop damage from insects like locusts, droughts, and other climate problems that will precipitate famine on a regional and perhaps global scale.
- Rising carbon dioxide (CO_2) levels will significantly reduce the nutritional value of plants, thus impairing the health of wildlife and the health of humans and their livestock.
- The permafrost is melting, causing millions of dollars of damage to roads, pipelines, and buildings and releasing vast amounts of methane

into the atmosphere, thereby worsening global warming. Craters are appearing in Siberia from exploding methane.

- As permafrost on beaches thaws and protective sea ice melts, shoreline erosion gets worse, forcing native villages in the Arctic to relocate further inland.
- Melting permafrost is thawing animal and human bodies, releasing dormant microbes. The 2016 anthrax outbreak in Siberia came from anthrax bacteria released from a melting reindeer carcass.

Air

- Climate change will make the air we breathe more polluted from ground-level ozone, wildfire smoke, allergens, and increased levels of dust and other particulate matter.
- Higher levels of CO_2, air pollution, and hotter air temperatures reduce human cognitive abilities.
- Dust storms are increasing globally, spreading dust, sand, pathogens, spores, agrochemicals, and toxic metals. Valley fever is an example of a fungal infection spread by wind that will likely spread further with warming temperatures.
- Tornadoes on average are becoming wider, stay on the ground longer, and occur more often as multiple funnels from the same storm. The area known as Tornado Alley is moving east into the Mississippi and Ohio River valleys.
- Hurricanes are getting bigger and more powerful, advance more slowly, and dump more rain than before. They also intensify faster and are reaching farther north.
- More destructive hurricanes mean higher costs for property insurance, disproportionately hurting low-income coastal communities.
- Heat waves kill more people in the United States than any other extreme weather hazard, especially harming the elderly, children, pregnant women, people with chronic health conditions, and those who work outside.
- A heat dome is a high-pressure system that traps hot air for long periods of time. Heat waves and heat domes are causing record high temperatures all over the world and are expected to increase in frequency and severity.
- A wet-bulb temperature is the air temperature at 100% humidity. Human beings cannot survive a wet-bulb temperature above 95°F (35°C). Parts of the world are starting to reach that critical

temperature. Extreme heat and humidity also kill livestock and wildlife.

Fire

- Wildfires are getting larger, causing more damage to people's homes, and wreaking havoc on ecosystems. Warming temperatures are making wildfires more intense, hotter, and more widespread. Fire seasons have been extended. Wildfires are no longer confined to forests and grasslands but burn right into our cities.
- Smoke from wildfires aggravates preexisting heart or lung disease, affecting people thousands of miles from the source of the smoke.

Water

- Melting glaciers and thermal expansion of seawater cause sea level rise, which will likely continue for centuries, overwhelming the world's coastal cities.
- By 2050 sea level rise will be 1 foot (30 cm) on the US East Coast and 18 inches (0.5 m) along the Gulf Coast. The National Oceanic and Atmospheric Administration (NOAA) projects a 5-foot to 8-foot maximum (1.5–2.4 m) global rise in sea level. If all the world's glaciers melt, ocean levels will rise more than 200 feet (61 m).
- The rate of sea level rise has more than doubled between 2000 and 2020.
- Just a 2-to-3-foot (0.6-to-0.9 m) rise in sea level will make most of the world's 2,150 barrier islands uninhabitable.
- King tides and other nuisance flooding are becoming more damaging to coastal communities as the sea rises. Nuisance flooding is a preview of a future with higher ocean levels.
- The sea absorbs about 30% of atmospheric CO_2. Seawater chemically interacts with CO_2 to form carbonic acid, which causes ocean acidification. Acidification is also happening in freshwater lakes.
- Ocean acidification dissolves the bodies of corals, shrimp, oyster larvae, clams, and other organisms made of calcium carbonate. At the same time, acidification makes it harder for organisms to find enough calcium to secrete their shells.
- The world's coral reefs may disappear due to warmer waters and ocean acidification. About 25% of all marine life occupies coral reefs.
- Marine heat waves (like the 2015 "Blob" off the US West Coast), extensive areas of overheated ocean water, similar to the heat domes

on land, are occurring globally. They can last for days, months, or even years.

- Marine heat waves decimate kelp forests, kill fish larvae and crabs, cause massive die-offs of seabirds, spread unwanted invasive species, and are bleaching the world's coral reefs.
- The snow crab and king crab fishery in Alaska was devastated by a combination of marine heat waves (and overfishing), causing the state to halt the fishery in the fall of 2022 in the hope of eventual recovery.
- As glaciers continue to melt in the fjords of Alaska and other Arctic regions, mountain slopes are losing the protective cover of ice, becoming dangerously unstable. Scientists expect some slopes (like those near Whittier, Alaska) will suddenly collapse into the sea and trigger deadly tsunamis, potentially hundreds of feet high.
- Meteotsunamis are tsunami-like waves formed from weather events on the world's lakes and oceans. They vary from 1 foot (30 cm) to nearly 20 feet (5.9 m) in height. More than 100 occur annually in the Great Lakes and on the US East Coast. They can cause fatalities and widespread damage. Climate change is expected to increase their frequency.
- A warming atmosphere holds more water (7% more for every 1°C [1.8°F] increase in temperature), triggering more extreme precipitation and flooding.
- Some of the worst flooding comes from atmospheric rivers, narrow bands of water vapor streaming across the sky. The flooding becomes catastrophic with the removal of all vegetation after a wildfire. With nothing to stabilize the slopes, massive mudslides occur, destroying houses, highways, and anything else in the way.
- Hotter temperatures are diminishing snowpack, drying soil and vegetation, and, along with changing rainfall patterns, causing droughts. Droughts are becoming larger, more frequent, more intense, and longer lasting.
- Droughts lead to water shortages, crop failures, food insecurity, increased potential for wildfires, dust storms, and topsoil loss. Droughts kill wildlife and decimate ecosystems.
- Drought in the American Southwest and northern Mexico is the worst in 1,200 years. In 2022 water levels in Lake Mead were at the lowest levels since the 1930s, threatening electricity for 1 million and drinking water for 25 million. Although atmospheric rivers and melting snowpack in 2023 replenished Lake Mead, warming temperatures and continued drought promise water-supply issues will go on for decades to come.

- Climate instability will lead to water shortages, diminishing water quality, collapsing ecosystems, fish kills from eutrophication, and declining freshwater and saltwater oxygen levels. As human population increases, armed conflict over water rights will happen, especially between countries using the same river for drinking and irrigation.

Space

- Heat waves and other climate disasters will render much of the land between 45° latitude north and 45° latitude south, especially near the tropics, virtually uninhabitable.
- Billions of climate refugees will relocate this century, either within their own national borders or in Europe and North America.
- Sea level rise alone will trigger 2 billion refugees by 2100.
- Climate havens are places where the effects of climate disasters are minimal and political will and infrastructure exist to handle future climate issues.
- Northern Europe (especially Scandinavia) and New Zealand are at the top of most climate haven lists. The Upper Midwest near the Great Lakes and parts of New England are considered the best climate havens in the United States.
- We must redesign cities to make them climate resilient. Some cities, like Miami, may have to be abandoned.
- Climate change is hurting our health. Hotter temperatures cause sleep deprivation, heat exhaustion, heatstroke, and heat death; increase cardiovascular and respiratory complications; and spread tropical diseases into temperate zones.
- The evacuation of people with chronic health issues requires extra planning and preparation.
- In response to global warming, plants and animals are evolving into new species and migrating toward the poles and to higher, cooler elevations. Some animals are changing their behavioral patterns to adapt to changing climate.
- Putting vast resources into colonizing Mars is a waste of time, money, and creative energy. We should first make Earth livable by reducing greenhouse gas emissions and halting climate change.
- In 2008 Ecuador was the first country to put the rights of Nature in its constitution. Several other countries have followed Ecuador's lead.

What to Do: Adapting to Climate Change

- **Behavior changes.** Eat more vegetables, drive less, and have fewer children. Stop building in areas at risk from wildfires, floods, sea level rise, storms, and so on. Move now to a safer area (climate haven). Evacuate promptly before extreme weather arrives. Consume less energy, food, water, and living space. Switch to crops appropriate for changing weather patterns. Wear light, loose-fitting clothes. No more business attire. Work outdoors in morning and evening or work from home during heat waves.
- **Nature-based adaptations.** Restore marshes and mangroves to increase spawning grounds for marine life and to make barriers against storm surge and rising seas. Plant trees in cities for shade. Create Nature preserves, wildlife corridors, vertical gardens, rain gardens, urban gardens, and green roofs. Practice afforestation. Build ice stupas and Warka Towers where feasible. Stop using chemicals on lawns, and do little watering. Reduce pesticide use.
- **Technological adaptations.** Hurricane and tornado shelters. Solar canals. Solar stills. Air filters for smoke pollution. Floating cities and floating houses. Underground cities. Desalination plants. Climate adaptation strategies from traditional cultures: wind towers for cooling, fire-resistant materials (adobe, brick, etc.). Assisted evolution. De-extinction. Geoengineering. For outdoor work, masks approved by the US Environmental Protection Agency (EPA).
- **Political adaptations.** Make evacuation plans for extreme weather events. Open all interstate lanes to expedite evacuation (hurricanes and wildfires). Pass appropriate building codes for new extreme climate events. Start making plans to house, feed, and assimilate the tens of millions of climate refugees. Legislate the rights of Nature. Elect science-literate leaders. Make national, regional and local climate change plans. Eliminate environmental injustice.
- **Education.** Fund academic, corporate, and NGO research for climate adaptation and mitigation solutions. Educate yourself and others about climate change, climate-related health issues, and so on. Teach environmental science (including the study of ecology, geology, and the evolution of life) at all levels of public education. Pay attention to weather reports, hurricane and tornado projections, wildfire risk levels, and air quality reports.

NEW IDEAS

The climate crisis has spurred creative thinking from engineers, architects, scientists, artists, writers, environmentalists, and philosophers, all pumping out new ideas to deal with climate change. Included in this mix are some interesting proposals for climate mitigation and adaptation:

Aerial seeding. Several companies have designed drones that fire "seed missiles" to regenerate forests and grasslands. Millions of mangrove trees have been successfully planted in Myanmar by drones built by the company Biocarbon Engineering. Ten drones can plant 100,000 trees a day, many times what can be done by hand. Similar drone seeding projects are underway in Dubai, Australia, and the fire-ravaged western United States (Peters 2017).

Assisted evolution. *Assisted evolution* is a general term for humans intervening to help Nature "evolve" to survive the rigors imposed by climate change. Examples include (1) genetically modifying an organism to enhance its survival chances and (2) exposing organisms to environmental stress to condition them to cope with future climate stresses. American scientists are using assisted evolution to help coral reefs survive in warmer, more acidic oceans (Kolbert 2021).

Assisted migration. In response to a warming climate, forests are migrating to cooler places, either upslope or toward the poles. Trees need to move 2 to 3 miles (3.2 to 4.8 km) each year to keep pace with climate change. On average, forests can only migrate perhaps 1,600 feet (488 m) a year, depending on the species. Assisted migration is when people transplant trees further north or upslope (further south in the Southern Hemisphere) to save a forest from succumbing to the drier, hotter conditions of the future. Some scientists are helping trees migrate by transplanting individual specimens that by chance are growing in tough conditions similar to those in the hotter world of the future.

Assisted migration may be the only way some species like Joshua trees (*Yucca brevifolia*), bristlecone pines (*Pinus longaeva*), and California's redwoods (*Sequoia sempervirens*) can survive. But physically moving an entire forest is harder than it seems. The symbiotic fungi wrapped around a tree's root system should also be transplanted, as well as the plants and animals that comprise the forest community. Even then there's no guarantee the forest will survive a rapidly changing climate (Ledig and Kitzmiller 1992; Markham 2021).

Biochar. Take any organic matter and bake it in a kiln in the absence of oxygen to produce a chemically stable charcoal-like substance called *biochar*. When buried in the soil, biochar sequesters atmospheric carbon for centuries while also improving soil fertility. Native Americans in Amazonia used biochar with "slash and char" farming for thousands of years, one reason the soils there are so fertile (Hawken 2017).

Carbon coin. A carbon coin is a kind of cryptocurrency like Bitcoin. The idea is explored in Kim Stanley Robinson's cli-fi classic *The Ministry for the Future* (2020). Carbon coins can be issued to anyone who can sequester significant amounts of carbon dioxide (CO_2) for a century or more by whatever means—planting trees, pulling carbon out of the air, rewilding, and the like. The carbon coins can be traded for other cryptocurrencies or exchanged for cash at a bank. This, of course, requires the banking system to back carbon coins. What about a climate adaptation coin (CAC coin) given to anyone who invents a new climate adaptation strategy that demonstrably saves lives?)

Carbon tax (aka carbon fee). A carbon tax is a tax levied on corporations that emit CO_2 or other greenhouse gases, forcing fossil fuel companies to pay for the true environmental costs of their products. Variants of the idea have been around at least since the 1970s. The tax revenues can be used to lower income taxes on the middle class, clean up toxic sites, invest in climate adaptation/mitigation policies, and so forth, or for charity.

Civilian Climate Corps. Inspired by the Civilian Conservation Corps (CCC) of the 1930s but focused on fixing the problems of climate change. This is an idea whose time has come (or will come, since the US Congress currently has shown little interest in funding the proposal). A Civilian Climate Corps would pay (mostly young) people to (1) plant trees, (2) help clean up after climate disasters, (3) deliver air filters to families downwind from wildfire smoke, (4) clear brush in national forests to reduce fire risk, (5) help build wildlife overpasses, (6) restore damaged ecosystems, (7) educate the public about climate issues, and so on. The list of possibilities goes on. The beauty of this idea is that the youth of the world are primed and ready to sign up.

De-extinction (resurrection biology, species revivalism). Scientists are actively trying to resurrect mammoths, the Pyrenean ibex, the passenger pigeon, the Tasmanian tiger, quaggas, aurochs, steppe bison, cave lions, and many other extinct animal and plant species through gene editing, selective breeding, cloning, and the like. The plan is to revive extinct species and reintroduce them into Nature. (See "Pleistocene Park," below.)

Some argue we have no right to "play God" with Nature. But the truth is we've been playing god with Nature ever since our ancestors departed Africa tens of thousands of years ago—exploiting ecosystems, destroying habitats, driving species to extinction, and triggering climate change. What is the alternative? Do nothing? Wave goodbye while species vanish into the night of time? Passively watch the wonder and beauty of the world disappear forever?

I am Lazarus, come from the dead, come back to tell you all...

"4 per 1,000" Initiative. Starting in France in 2015, the International "4 per 1,000" Initiative (n.d.) is an attempt to get farmers to help mitigate climate change and enhance food security by increasing carbon storage in agricultural soils by 0.4% every year. If done globally, the initiative would reverse the annual 40 billion (36 billion metric tons) or so tons of carbon humans put in the atmosphere. This would boost soil fertility and reduce soil degradation (now threatening 40% of the Earth's land surface). The initiative proposes sequestering carbon by (1) planting hedges and trees along field boundaries, (2) restoring degraded farmland in arid regions, (3) practicing sustainable agriculture (low tillage, agroecology, agroforestry) and sustainable grazing, and (4) subsidizing the education of farmers while promoting research to find more sustainable farming techniques. One hundred sixty countries signed the initiative. The United States did not (Rumpel et al. 2020).

Geoengineering. There are two principal ways to engineer a cooler climate: (1) reflect sunlight away from the Earth or (2) remove CO_2 from the atmosphere. The first approach has inspired some serious, but bizarre, methods for reflecting sunlight: depositing inflatable bubbles in space (MIT 2022), filling the sea with microbubbles, pumping water that freezes on sea ice to increase its albedo (reflectivity), seeding clouds over the ocean to make them brighter (Smedley 2019), or dropping sulfates in the upper atmosphere. By contrast, carbon removal is more prosaic. Countries like the

United States, Iceland, and others are building huge factories to suck carbon out of the air and pump it deep into the ground, a method that is currently prohibitively expensive on a scale large enough to cool the atmosphere. Of course, planting trees does the same thing, sequestering carbon during the lifetime of the trees, which, depending on the species, could be decades, centuries, or millennia.

Geoengineering in its various guises is potentially a risky venture. But many climate scientists agree it should be included in the mix of remedies to counter climate change. Moreover, even if the United States and the European Union ban geoengineering, other countries will try to engineer a cooler climate. Some countries in Asia or Africa may face a hard choice: either geoengineer a livable climate or face national extinction.

Kelp farming. When a plant dies, it is decomposed by microbes. All the oxygen produced over the course of the plant's life gets consumed by the microbes. Thus, living plants (trees, algae, etc.) over the long run produce no oxygen . . . unless a plant falls into sediment and becomes fossilized. Then the carbon in the plant's cells is permanently sequestered, and the oxygen that normally would have been chemically used up with decomposition remains in the atmosphere. Over millions of years, enough photosynthesizers get fossilized that atmospheric oxygen levels stay more or less the same. This is the concept behind kelp harvesting. Harvest kelp, lots of kelp. Drop it to the seafloor where it won't decompose, thus sequestering the carbon from its cells. Oxygen levels go up. Carbon dioxide (CO_2) levels in the air and sea drop, reducing global warming and diminishing ocean acidification. Have you thanked a brown fossil today? (Bever 2021).

Lab-grown meat. Already legal in Singapore and just approved by the US government, lab-grown meat could help reduce methane emissions from cattle and alleviate food shortages expected from our fast-warming climate.

Pleistocene park. In northeastern Siberia the Russian geophysicist Sergey Zimov is trying to reconstruct the grassland ecosystem of a Pleistocene landscape. He is reintroducing Ice Age species like musk oxen, bison, yaks, and possibly in the future mammoths (if and when they are brought back from extinction) (see figure 87). The goal of the project is to fight climate change by keeping the permafrost frozen. Apparently the Pleistocene herds used to dig through the snow with their hooves to eat the grass that once covered the Siberian steppes. This removed the snow cover, thereby eliminating snow's insulating properties. Because the permafrost was exposed to colder air temperatures, the ground remained colder, preventing

Figure 87. Mammoth. Drawing by Charles Pilkey.

melting of the permafrost (i.e., it remained permanently frozen). This process would reduce the contribution of greenhouse gases from melting permafrost and also prevent the destabilizing effect on buildings. If Pleistocene Park succeeds, it may inspire similar projects elsewhere.

Robot firefighters. Robots have already been built to fight urban fires. Now they are being designed to locate and control wildfires as well (Little 2021).

Space-based solar power. This is a technology the Japanese and Russians are heavily investing in. The idea is to place solar panels in space and beam electricity down to the Earth's surface in the form of microwaves that can be converted back to conventional electricity (Wood 2014).

Sustainable capitalism (green capitalism, regenerative capitalism, eco-capitalism). Capitalism is an economic system that (directly or indirectly) commodifies Nature and destroys ecosystems to create corporate wealth. Over time that wealth becomes concentrated in an increasingly smaller percentage of the world's population. The rich get richer; the middle class and poor get poorer. The historical Marxist-based alternatives have been unmitigated disasters, spawning barbarous dictatorships that have left a wake of horrendous environmental destruction. Sustainable capitalism is an attempt to preserve and regenerate natural resources instead of expending them for profit, to keep the best of capitalism, its creativity and affluence, while maintaining biodiversity.

THE OLD CAPITALIST SPEAKS: *How can I maximize my own personal wealth?*

THE NEW CAPITALIST SPEAKS: *How can I make wealth that benefits Nature, self, and society?*

Zen and the art of biospheric maintenance. Al Gore and others have argued that a spiritual transformation of humanity is needed to make us better stewards of the Earth. Graphs, scientific papers, reason . . . are not enough to sway the masses. What is needed, or so the argument goes, is a new religion, a Gaia-centered worldview that promotes the rights of Nature (see "The Biosphere" in "Space") and instills reverence for Pachamama (Mother Earth). The world's major religions are hopelessly anthropocentric systems, originating for the most part in desert cultures far removed from modern industrial civilizations. Nowhere in the Sermon on the Mount does Christ mention greenhouse gases. Nor is preventing ozone depletion a pillar of Islam. Even the ahimsa (nonviolence or do not harm life) of Hinduism, Budhism, and Jainism is practiced to avoid negative karmic consequences for humans.

Some have tried to reorganize old belief systems, taking relevant parts of different religions—the ahimsa of Hinduism, Indigenous spirituality, the turn-the-other-cheek ethos of Christian ethics, the conservation of resources mentioned in the Koran, and so forth, reshaping them into a new, syncretic, Nature-centered worldview. But this requires some serious mental gymnastics, rather like fitting a square theological peg into a round spiritual hole.

Religions are started somewhere by someone. They grow organically from the fertile grounds of a particular historical and cultural context. Could it be that right now a modern-day Moses is wandering an industrial wilderness, pondering how best to bring her people home to the Promised Land of ecological harmony?

Namaste. Amen. Shalom . . .

NEW DEVELOPMENTS

Nearly every day new research or an extreme weather event of one kind or another reminds us how fast the world's climate is changing. Here is a skeletal summary of some important concepts or events that have occurred during the publication of *Escaping Nature*:

The latest United Nations report. The 2023 report from the United Nations International Panel on Climate Change "declares the science unequivocal and warned that even with urgent action we will face a dramatic uptick in catastrophic events from droughts to floods to fires" (IPCC 2023).

Doomsday glacier. West Antarctica's Florida-sized Thwaites Glacier is retreating faster than expected. If it collapses, the ice behind it could also collapse, raising ocean levels by as much as 10 feet (3.5 m) (Lea 2023).

Antarctic meltdown. According to a CNN report, 90% of Antarctica's sea ice disappeared between 2014 and 2023 (CNN 2023).

Melting mountain glaciers. Hinman Glacier (Mount Rainier's largest) completely disappeared in 2023. Nearly half the world's mountain glaciers are expected to do the same by 2100 (Ramirez 2023a).

The summer of '23. NASA announced Earth's summer in 2023 was the hottest ever recorded. Cities around the world experienced record-breaking high temperatures. The ocean off Florida exceeded 101°F (38°C), possibly the world's hottest ever recorded sea temperature. Spurred by the heat, wildfires burned more than 5% of Canada's forests. Coral bleaching, deadly floods, punishing heatwaves, extreme weather, rising CO_2 levels . . . hard times for the denizens of Pachamama.

Massive floods in New Zealand. On January 27, 2023, Auckland received a summer's worth of rain in 24 hours, unleashing what some have called the "biggest climate event" in New Zealand's history. Flooding worsened a few days later when Gabrielle (the costliest cyclone ever in the Southern Hemisphere) brushed the North Island (McLay 2023).

Deadly floods in Libya. From September 10–11, 2023, heavy rains from Storm Daniel caused two dams to burst near the city of Derna. The ensuing flood killed between 4,000 and 10,000 people. Mediterranean storms like Daniel that resemble hurricanes in shape and wind speed (known colloquially as *medicanes*) are expected to become more potent with a warming climate.

Planetary boundaries. Scientists recognize nine interconnected planetary boundaries within which humanity can safely live. These boundaries are climate change, biodiversity loss, ocean acidification, freshwater use, land use change, nitrogen and phosphorus cycles, atmospheric aerosols, ozone depletion, and novel entities (pollution, microplastics, etc.). In 2023, researchers concluded that six of those boundaries have been dangerously transgressed. No one knows what this means for our future... but it can't be good.

American Climate Corps. In 2023, President Joe Biden founded the American Climate Corps. Inspired by Franklin D. Roosevelt's Depression-era Civilian Conservation Corps, Biden's climate corps will train and pay 20,000 young people to restore wetlands, manage forests, and help with climate disaster relief.

BUG-OUT BAGS

"PREPAREDNESS PERFECTS RESPONSE"

Experience has shown that preparation for escaping an impending fire, hurricane, flood, tsunami, or heat wave is a good idea. It may even save your life and the lives of your loved ones. The contents of a good Bug-Out Bag will vary depending on where you are, the kind of climate event, the season of the year, the needs of your family, your plans for a rendezvous, and who is with you (kids, pets, neighbors). Gather what you need now and push past the inertia that prevents us from preparing for disasters. Decide how many days you need to be away—it should be at least three days. Make an evacuation plan. Have a backup plan in case things go awry. What if the car won't start?

Use airtight plastic bags and put your supplies in an easy-to-carry container (a sports bag, backpack, suitcase, cooler, or similar container). Buy a battery-powered, hand-crank emergency weather radio with tone alert. Think about where to store your bags for a quick getaway—in your car? Your home? Your office?

Food

- **Water.** This is the most essential survival item. You can live for weeks without food but only for days without water.
- **Snacks.** Dried foods, juice, crackers, nuts, trail mix.
- **Canned goods and a can opener.** A Swiss army knife (or equivalent) would be handy.
- **Eating utensils**

Personal

- **Documents.** Insurance policies, passports, birth certificates, social security cards, etc. (all in a waterproof, portable container).
- **Maps**
- **First aid kit**

- **Pencil, pen and paper**
- **Cash, credit cards, checks**
- **Cell phone and chargers.** Don't assume cell service will be available.
- **Light source.** Flashlights and batteries, LED headlamp, candles, matches.
- **Personal hygiene items.** Toilet paper, toothpaste, toothbrush, bug repellent, face masks, prescription medications, period care, sunscreen, soap, hand sanitizer, paper towels, disinfecting wipes.

Other

- **Full tank of gas**
- **Plastic sheeting and duct tape**. For sheltering in place.
- **12-Volt tire inflator.** One that connects to your car's power outlet.
- **Tools.** Pocket knife, wrench, hatchet, basic tool kit, tire jack.
- **Blankets or sleeping bags**
- **Pet supplies, pet carrier**
- **Spare clothes**
- **Tent**
- **Morale boosters.** Books, games, puzzles, activities for children.

Maintenance

- **Update your bag.** Reexamine your needs every year and update accordingly.
- **Do evacuation practice runs.**
- **Post evacuation plans on refrigerator for quick reference.**

TO LEARN MORE

RESOURCES IN PRINT AND ON SCREEN

The Best of the Best

The Sixth Extinction: An Unnatural History, by Elizabeth Kolbert. This 2014 Pulitzer Prize winner chronicles the history of human-induced extinctions from prehistoric times to the present. Kolbert also wrote *Under a White Sky,* about ways to help Nature survive climate change and biodiversity loss through assisted migration, genetic modification, geoengineering, and other interventions.

The Uninhabitable Earth: Life after Warming, by David Wallace-Wells. Sometimes called the *Silent Spring* of our times, this is the best summary of the complex problems posed by our changing climate.

Half-Earth: Our Planet's Fight for Life, by Edward O. Wilson. Wilson proposes setting aside half the Earth and oceans as Nature preserves as a way to preserve biodiversity.

The Best of the Rest

The Water Will Come: Rising Seas, Sinking Cities, and the Remaking of the Civilized World, by Jeff Goodell. A very readable summary of sea level rise and its implications for civilization.

Drawdown: The Most Comprehensive Plan Ever to Reverse Global Warming, edited by Paul Hawken. Written by scientists at the forefront of devising strategies to mitigate climate change.

Global Climate Change: A Primer, by Orrin Pilkey and Keith Pilkey. A general discussion of climate change and the phenomenon of climate change deniers/minimizers.

How to Prepare for Climate Change: A Practical Guide to Surviving the Chaos, by David Pogue. Six-hundred-plus pages of stratagems for getting through the tough times ahead.

How to Avoid a Climate Disaster, by Bill Gates. What technologies must we develop to get to zero emissions?

Sea Level Rise: A Slow Tsunami on America's Shores, by Keith Pilkey and Orrin Pilkey. The impact of sea lea level rise on US shorelines and how various groups such as the Inupiat will be affected.

Retreat from a Rising Sea: Hard Choices in an Age of Climate Change, by Orrin Pilkey, Linda Pilkey, and Keith Pilkey. Emphasizes retreating from a shoreline as the best response to sea level rise.

The Icy Planet: Saving Earth's Refrigerator, by Colin Summerhayes. The history and fate of the Earth's cryosphere, how ice regulates Earth's climate, and what will happen in the future as Earth's glaciers melt.

Our Fragile Moment, by Michael Mann. What the geologic history of climate change can tell us about the future of climate change. Conclusions based on the latest computer modeling.

Environmental Classics

Desert Solitaire, by Edward Abbey. In his humorous, irascible style Abbey writes about his experiences as a park ranger in Utah while also thinking about the friction between modern technological society and Nature. Also wrote *The Monkey Wrench Gang,* a story about a group of concerned citizens who employed some rather unconventional strategies to protest environmental degradation.

Silent Spring, by Rachel Carson. The 1962 classic that launched the modern environmental movement.

Sand County Almanac, by Aldo Leopold. The book that proposed the idea of a land ethic, the precursor to the rights of Nature.

Walden; or Life in the Woods, by Henry David Thoreau. Ruminations about Nature, society, and the human condition. Inspiration for Rachel Carson and pretty much everybody else. Thoreau also wrote *Civil Disobedience,* the book that inspired Gandhi, Nelson Mandela, Martin Luther King, and

Cesar Chavez to lead protest movements against social injustice (all four men read the book while in jail).

Cli-fi (Science fiction about climate change)

The Ministry for the Future, by Kim Stanley Robinson. A not-too-distant future that unfolds after a catastrophic heat wave in India. The author proposes thoughtful solutions to our climate crisis (including the idea of a carbon coin). Other cli-fi works by Robinson: *New York 2140* and *2312*.

The Drowned Cities, by Paolo Bacigalupi. The best in the Ship Breaker trilogy (*Ship Breaker, The Drowned Cities, Tool of War*) dealing with sea level rise and genetic engineering. Other cli-fi works: *The Water Knife* and *The Windup Girl*.

The Parable of the Sower, by Octavia Butler. A dystopian future of climate change and extreme disparity of wealth.

Oryx and Crake, by Margaret Atwood. A dark parable of the risks of climate change and genetic engineering.

The Overstory, by Richard Powers. Stories centered on our relationship with trees and the need to preserve the world's forests.

Children's Books

The Most Important Comic Book on Earth. One hundred twenty visual stories by artists, writers, musicians, environmentalists, and other creative people to save the world.

The Magic Dolphin, by Charles Pilkey. A genetically engineered "super" dolphin teaches kids about global warming, sea level rise, and other marine environmental issues.

A Hot Planet Needs Cool Kids, by Julie Hall. The causes and effects of climate change and what people can do about it.

The Magic School Bus and the Climate Challenge, by Joanna Cole. Another in the popular Magic School Bus series.

Documentary Films

An Inconvenient Truth (2006), directed by Davis Guggenheim. Al Gore's efforts to raise awareness about global warming.

An Inconvenient Sequel: Truth to Power (2017), directed by Bonnie Cohen and Jon Shenk. The follow-up to *An Inconvenient Truth* one decade on.

Before the Flood (2016), directed by Fisher Stevens for National Geographic. Narrated by Leonardo DiCaprio, who travels the world to see the effects of climate change.

Chasing Ice (2012), directed by Jeff Orlowski. Beautiful time-lapse photography of retreating glaciers showing the impact of climate change on the planet's ice.

Cowspiracy: The Sustainability Secret (2014), directed by Kip Anderson and Keegan Kuhn. The connections between cows and methane emissions.

Ice on Fire (2019), directed by Leila Conners. Narrated by Leonardo DiCaprio. Cites the evidence that climate change is happening now and explores ways to resolve the climate crisis.

Kiss the Ground (2013), directed by Josh and Rebecca Tickell. Narrated by Woody Harrelson. How regenerative farming, low tillage, ground cover, and other techniques can replenish vanishing topsoil, sequester carbon, and reduce global warming.

Television

Apple TV's *Extrapolations*. An intellectually dense drama about future climate change coupled with developments in AI.

EarthxTV. A network dedicated to promoting environmental sustainability. Excellent series on climate change adaptations, sea level rise, and green cities.

Link TV. Environmental issues discussed on *Bioneers*, *Earth Focus*, and other episodes.

REFERENCES AND ADDITIONAL SOURCES

Preface

Smith, Charlie. 2021. "Climate Stability Becomes a Relic of the Past in B.C. Due to Loss of Hydrologic Stationarity." *Georgia Straight*, Vancouver Free Press, December 1, 2021. https://www.straight.com/news/climate-stability-becomes-a-relic-of-past-in-bc-due-to-loss-of-hydrologic-stationarity.

Introduction

Guterres, António. 2022. "UN Secretary General António Guterres at the Opening Ceremony of the World Leaders Summit—#COP27." YouTube video, UN Climate Change, November 7, 2022. https://www.youtube.com/watch?v=YAVgd5XsvbE.

Earth

THE LESSONS OF GEOLOGIC TIME

Brannen, Peter. 2017. *The Ends of the World: Volcanic Apocalypses, Lethal Oceans, and Our Quest to Understand Earth's Past Mass Extinctions.* New York: Ecco/HarperCollins.

Graham, Joseph, William Newman, and John Stacy. 2008. "The Geologic Time Spiral—a Path to the Past." US Geological Survey (USGS) General Information Publication 58. Updated September 2008. https://doi.org/10.3133/gip58.

Kolbert, Elizabeth. 2014. *The Sixth Extinction: An Unnatural History*. New York: Henry Holt.

Wilson, Edward O. 2016. *Half-Earth: Our Planet's Fight for Life*. New York: Liveright/W. W. Norton.

THE 2021 UNITED NATIONS CLIMATE REPORT

Alayza, Natalia, Preety Bhandari, David Burns, Nathan Cogswell, Kiyomi de Zoysa, Mario Finch, Taryn Fransen, et al. 2022. "COP27: Key Takeaways and What's Next. " World Resources Institute, December 8, 2022. https://www.wri.org/insights/cop27-key-outcomes-un-climate-talks-sharm-el-sheikh.

Arias, Paola A., Nicolas Bellouin, Erika Coppola, Richard G. Jones, Gerhard Krinner, Jochem Marotzke, Vaishali Naik, et al. 2021. "Technical Summary." In *Climate Change 2021: The Physical Science Basis; Contribution of Working Group I to the Sixth Assessment Report of the Intergovernmental Panel on Climate Change*, edited by Valérie Masson-Delmotte, Panmao Zhai, Anna Pirani, Sarah L. Connors, Clotilde Péan, Sophie Berger, Nada Caud, et al., 33–144. Cambridge: Cambridge University Press. https://doi.org/10.1017/9781009157896.002.

Berardelli, Jeff. 2021. "Major U.N. Climate Report Warns of 'Extreme' and 'Unprecedented' Impacts." CBS News, August 10, 2021. https://www.cbsnews.com/news/climate-change-impact-warning-report-united-nations-intergovernmental-panel-ipcc-code-red-humanity/.

IPCC (Intergovernmental Panel on Climate Change). 2021a. *Climate Change 2021: The Physical Science Basis; Contribution of Working Group I to the Sixth Assessment Report of the Intergovernmental Panel on Climate Change*, edited by Valérie Masson-Delmotte, Panmao Zhai, Anna Pirani, Sarah L. Connors, Clotilde Péan, Sophie Berger, Nada Caud, et al. Cambridge: Cambridge University Press. https://www.ipcc.ch/report/ar6/wg1/#SPM.

IPCC (Intergovernmental Panel on Climate Change). 2021b. "Summary for Policymakers." In *Climate Change 2021: The Physical Science Basis; Contribution of Working Group I to the Sixth Assessment Report of the Intergovernmental Panel on Climate Change*, edited by Valérie Masson-Delmotte, Panmao Zhai, Anna Pirani, Sarah L. Connors, Clotilde Péan, Sophie Berger, Nada Caud, et al., 3–32. Cambridge: Cambridge University Press. https://www.ipcc.ch/report/sixth-assessment-report-working-group-i/. https://doi.org/10.1017/9781009157896.001.

IPCC (Intergovernmental Panel on Climate Change). 2022. *Climate Change 2022: Impacts, Adaptation and Vulnerability. Contribution of Working Group II to the Sixth Assessment Report of the Intergovernmental Panel on Climate Change*, edited by Hans-Otto Pörtner, Debra C. Roberts, Melinda M. B. Tignor, Elvira Poloczanska, Katja Mintenbeck, Andrés Alegría, Marlies Craig, et al. Cambridge: Cambridge University Press. https://doi.org/10.1017/9781009325844.

Januta, Andrea. 2021. "Key Takeways from the U.N. Climate Panel's Report." Reuters, August 9, 2021. https://www.reuters.com/business/environment/key-takeaways-un-climate-panels-report-2021-08-09/.

Johnson, Boris. 2021. "Boris Johnson COP26 Climate Summit Glasgow Speech Transcript." Revised November 1, 2021. https://www.rev.com/blog/transcripts/boris-johnson-cop26-climate-summit-glasgow-speech-transcript.

UNEP (United Nations Environment Programme). 2022. *Emissions Gap Report 2022: The Closing Window—Climate Crisis Calls for Rapid Transformation of Societies*. Nairobi: UN Environment Programme. https://www.unep.org/emissions-gap-report-2022.

FAMINE

"Climate Change Makes Pests Move North from the Tropics—Study." 2013. *Guardian*, September 2, 2013. https://www.theguardian.com/environment/2013/sep/02/climate-change-crop-pests.

Deutsch, Curtis A., Joshua J. Tewksbury, Michelle Tigchelaar, David S. Battisti, Scott C. Merrill, Raymond B. Huey, and Rosamond L. Naylor. 2018. "Increase to Crop Losses to Insect Pests in a Warming Climate." *Science* 361 (6405): 916–19. https://doi.org/10.1126/science.aat3466.

Evich, Helena Bottemiller. 2017. "The Great Nutrient Collapse." *Politico*, September 13, 2017. https://www.politico.com/agenda/story/2017/09/13/food-nutrients-carbon-dioxide-000511/.

Hasell, Joe, and Max Roser. 2017. "Famines." Our World in Data. First published in 2013, substantive revision on December 7, 2017. https://ourworldindata.org/famines.

International Food Policy Research Institute. 2019. "Rising CO_2, Climate Change Projected to Reduce Availability of Nutrients Worldwide: Protein, Iron, Zinc to Be 19.5%, 14.4%, and 14.6% Lower, Respectively, Than without Climate Change." ScienceDaily, July 18, 2019. https://www.sciencedaily.com/releases/2019/07/190718085308.htm.

Njagi, David. 2020. "The Biblical Locust Plagues of 2020." BBC, August 6, 2020. https://www.bbc.com/future/article/20200806-the-biblical-east-african-locust-plagues-of-2020.

NOAA (National Oceanic and Atmospheric Administration). 2020. "NOAA Teams with United Nations to Create Locust-Tracking Application: Model Helps Warn of Worst Desert Locust Swarms in a Quarter Century." NOAA Research News. May 7, 2020. https://research.noaa.gov/article/ArtMID/587/ArticleID/2620/NOAA-teams-with-the-United-Nations-to-create-locust-tracking-application.

Ryckman, Lisa Levitt. 1999. "The Great Locust Mystery: Grasshoppers That Ate the West Became Extinct." *Denver Rocky Mountain News*, June 22, 1999. https://web.archive.org/web/20090601171826/http://enver.rockymountainnews.com/millennium/0622mile.shtml.
"Solar Drying." 2019. *Energypedia*. Last edited June 17, 2019. https://energypedia.info/wiki/Solar_Drying. Originally published by GIZ HERA and based on the publication "Productive Use of Thermal Energy—an Overview of Technology Options and Approaches for Promotion," published by GIZ Programme "Poverty-Oriented Basic Energy Services" and European Union Energy Initiative Partnership Dialogue Facility.
UN (United Nations). 2021. "Madagascar: Severe Drought Could Spur World's First Climate Change Famine." UN News: Global Perspective, Human Stories. October 21, 2021. https://news.un.org/en/story/2021/10/1103712.
UNEP (United Nations Environment Programme). 2020. "Locust Swarms and Climate Change." Story, Climate Action, February 6, 2020. https://www.unep.org/news-and-stories/story/locust-swarms-and-climate-change.
Wallace-Wells, David. 2020. *The Uninhabitable Earth: Life after Warming.* New York: Tim Duggan Books/Penguin Random House.
WFP (World Food Programme). 2021. "Famine Prevention." World Food Programme. Accessed August 24, 2023. https://www.wfp.org/famine-prevention.

PERMAFROST

Aridi, Rasha. 2021. "As Arctic Sea Ice Retreats, Orcas Are on the Move, Spurring Changes in the Food Chain." *Smithsonian Magazine*, December 3, 2021. https://www.smithsonianmag.com/smart-news/orcas-move-in-as-arctic-sea-ice-retreats-spurring-changes-in-the-food-chain-180979163/.
Boyce, Rod. 2021. "Researcher Finds Alaska's Arctic Coastal Towns Face Extensive Inundation." News and Information, University of Alaska Fairbanks. December 17, 2021. https://uaf.edu/news/researcher-finds-alaskas-arctic-coastal-towns-face-extensive-inundation.php.
Buldovicz, Sergey N., Vanda Z. Khilimonyuk, Andrey Y. Bychkdov, Evgeny N. Ospennikov, Sergey A. Vorobyev, Aleksey Y. Gunar, Evgeny I. Gorshkov, et al. 2018. "Cryovolcanism on the Earth: Origin of a Spectacular Crater in the Yamal Peninsula (Russia)." *Scientific Reports* 8:13534. https://www.nature.com/articles/s41598-018-31858-9.
Cheng, Feng, Carmala Garzione, Xiangzhong Li, Ulrich Salzmann, Florian Schwarz, Alan M. Haywood, Julia Tindall, et al. 2022. "Alpine Perma-

frost Could Account for a Quarter of Thawed Carbon Based on Plio-Pleistocene Paleoclimate Analogue." *Nature Communications* 13:1329. https://doi.org/10.1038/s41467-022-29011-2.

Doucleff, Michaeleen. 2019. "Storytelling Instead of Scolding: Inuit Say It Makes Their Children More Cool-Headed." *All Things Considered*, NPR, March 4, 2019. https://www.npr.org/2019/03/04/689925669/storytelling-instead-of-scolding-inuit-say-it-makes-their-children-more-cool-hea.

Farquharson, Louise Melanie, Dmitry Nicolsky, Anna M. Irrgang, Vladimir E. Romanovsky, Benjamin M. Jones, Ming Xiao, Min Liew, and Ann Gibbs. 2021. "C25F-0883—Permafrost Thaw and Coastal Erosion between 1950 and 2100 at Three Coastal Communities in Arctic Alaska, Part Observations and Future Projections." Paper presented at the American Geophysical Union Fall Meeting, New Orleans, LA, December 14, 2021. https://agu.confex.com/agu/fm21/meetingapp.cgi/Paper/989649.

Gray, Richard. 2020. "The Mystery of Siberia's Exploding Craters." BBC, November 30, 2020. https://www.bbc.com/future/article/20201130-climate-change-the-mystery-of-siberias-explosive-craters.

Hasemyer, David. 2021. "Unleashed by Warming, Underground Debris Fields Threaten to 'Crush' Alaska's Dalton Highway and the Alaska Pipeline." Inside Climate News, December 20, 2021. https://insideclimatenews.org/news/20122021/alaska-frozen-debris-lobes-dalton-highway-pipeline-climate-change/.

Joint Ocean Commission Initiative. 2017. "Arctic Town Votes to Relocate to Avoid Rising Seas." Ocean Action Agenda. March 2017. https://oceanactionagenda.org/story/arctic-town-climate-change/.

Jones, B. M., A. M. Irrgang, L. M. Farquharson, H. Lantuit, D. Whalen, S. Ogorodov, M. Grigoriev, et al. 2020. "Arctic Report Card 2020: Coastal Permafrost Erosion." Arctic Program, National Oceanic and Atmospheric Administration. https://doi.org/10.25923/e47w-dw52.

Jones, Verona. 2020. "Alaska's Qalupalik." Coffee House Writers. August 18, 2020. https://coffeehousewriters.com/alaskas-qalupalik/.

Kin, Greg. 2019. "Residents of an Eroded Alaskan Village Are Pioneering a New One, in Phases." *All Things Considered*, NPR, November 2, 2019. https://www.npr.org/2019/11/02/774791091/residents-of-an-eroded-alaskan-village-are-pioneering-a-new-one-in-phases.

Mason, Owen K., James W. Jordan, and Lawrence Plug. 1995. "Late Holocene Storm and Sea Level History in the Southern Chukchi Sea." In "Holocene Cycles: Climate, Sea Levels, and Sedimentation," edited by C. W. Finkle. *Journal of Coastal Research*, special issue no. 17, 173–80. https://www.jstor.org/stable/25735641.

Mele, Christopher, and Daniel Victor. 2016. "Reeling from the Effects of Climate, Alaskan Village Votes to Relocate." *New York Times*, August 19, 2016. https://www.nytimes.com/2016/08/20/us/shishmaref-alaska -elocate-vote-climate-change.html.

National Park Service. 2015. "Bering Land Bridge National Preserve, Alaska: Shishmaref." Last updated May 1, 2015. https://www.nps.gov /bela/planyourvisit/shishmaref.htm.

Nature Portfolio. 2022. "Permafrost in a Warming World." *Nature*, January 11, 2022. https://www.nature.com/collections/ababhihdce.

NSF Public Affairs. 2022. "Fast-Melting Alpine Permafrost May Contribute to Rising Global Temperatures." National Science Foundation Research News. April 18, 2022. https://new.nsf.gov/news/fast-melting-alpine -permafrost-may-contribute.

Osborne, Hannah. 2017. "Massive Craters from Methane Explosions Discovered in Arctic Ocean Where Ice Melted." *Newsweek*, June 1, 2017. https://www.newsweek.com/hundreds-craters-methane-explosions -seafloor-arctic-norway-russia-619068.

Paoli, Julia. 2015. "30,000-Year-Old Virus Found in Siberian Permafrost." *Viruses 101* (blog), Scitable, September 19, 2015. https://www.nature.com /scitable/blog/viruses101/30000yearold_virus_found_in_siberian/.

UNEP (United Nations Environment Programme). n.d. "Methane." United Nations Environment Programme Climate and Clean Air Coalition. Accessed June 12, 2022. https://www.ccacoalition.org/en/slcps/methane.

U.S. Climate Resilience Toolkit. 2021. "Relocating Kivalina." Last modified August 9, 2021. https://toolkit.climate.gov/case-studies/relocating -kivalina.

USGS (US Geological Survey). 2018. "Gas Hydrates — Primer." Woods Hole Coastal and Marine Science Center. March 19, 2018. https://www.usgs .gov/centers/whcmsc/science/gas-hydrates-primer.

Wallace-Wells, David. 2020. *The Uninhabitable Earth: Life after Warming*. New York: Tim Duggan Books/Penguin Random House.

Wilkerson, Jordan. 2021. "How Much Worse Will Thawing Arctic Permafrost Make Climate Change?" *Scientific American*, August 11, 2021. https://www.scientificamerican.com/article/how-much-worse-will -thawing-arctic-permafrost-make-climate-change/.

Air

HURRICANES

AJAT Philippines, Inc. 2021. "Dome Building Technology." Accessed April 21, 2022. https://ajatphilippinesinc.com/domebuildingtechnology.php.

Brown, Daniel P. 2021. *Hurricane Larry.* Tropical Cyclone Report AL 122021. National Hurricane Center, NOAA (National Oceanic and Atmospheric Administration), December 16, 2021. https://www.nhc.noaa.gov/data/tcr/AL122021_Larry.pdf.

Budryk, Zack, and Rachel Frazin. 2021. "Scientists Detail Role of Climate Change in Ida's Intensity." *The Hill*, August 31, 2021. https://thehill.com/policy/energy-environment/570087-scientists-detail-role-of-climate-change-in-idas-intensity.

Challoner, Jack. 2014. *Hurricane and Tornado.* London: Dorling Kindersley.

Editorial Board. 2021. "Hurricane Ida Drives Home Hellish Battle with Climate Change." Opinion, *USA Today*, August 31, 2021. https://www.usatoday.com/story/opinion/todaysdebate/2021/08/31/hurricane-ida-climate-change-hellish-summer/5655806001/.

Expatriate Group. 2018. "Expat Safety Tips: What to Do When a Hurricane Hits." Expatriate Healthcare. March 18, 2018. https://www.expatriatehealthcare.comexpat-safety-tips-what-to-do-when-a-hurricane-hits/.

Landsea, Chris. 2016. "Total and Average Number of Tropical Cyclones by Month (1851–2017)." National Hurricane Center, Atlantic Oceanographic and Meteorological Laboratory, NOAA (National Oceanic and Atmospheric Administration). Archived from the original on June 1, 2016. https://web.archive.org/web/20180901052811/http://www.aoml.noaa.gov/hrd/tcfaq/E17.html .

Larson, Erik. 1999. *Isaac's Storm: A Man, a Time, and the Deadliest Hurricane in History.* New York: Random House.

Mendoza, Jessica. 2015. "Could 'Dome Homes' Be the Future of Reducing Disaster Risk?" *Christian Science Monitor*, May 27, 2015. https://www.csmonitor.com/Technology/2015/0527/Could-dome-homes-be-the-future-of-reducing-disaster-risk.

Monolithic Dome Institute. n.d. "Monolithic Dome Homes, Safe Rooms, Schools, Storages, Gyms and More." Accessed April 21, 2022. https://www.monolithic.org/.

National Weather Service. n.d. "Hurricane Harvey and Its Impacts on Southeast Texas: August 25–29, 2017." NOAA (National Oceanic and Atmospheric Administration). Accessed September 22, 2022. https://www.weather.gov/hgx/hurricaneharvey.

NOAA (National Oceanic and Atmospheric Administration). 2021a. "Hurricanes: Frequently Asked Questions." Atlantic Oceanographic and Meteorological Laboratory. Revised June 21, 2021. https://www.aoml.noaa.gov/hrd-faq/#1569507388495-a5aa91bb-254c.

NOAA (National Oceanic and Atmospheric Administration). 2021b. "A World First: Ocean Drone Captures Video from inside a Hurricane." Atlantic Oceanographic and Meteorological Laboratory. September 30, 2021. https://www.aoml.noaa.gov/news/saildrone-captures-video-from-inside-hurricane/.

Parker, Freda. 2008. "A Village Grows: Progress in New Ngelepen, Indonesia." Monolithic Dome Institute. December 1, 2008. https://www.monolithic.org/new-ngelepen/a-village-grows-progress-in-new-ngelepen-indonesia.

Penny, Veronica. 2021. "What We Know about Climate Change and Hurricanes." *New York Times*, August 29, 2021. https://www.nytimes.com/2021/08/29/climate/climate-change-hurricanes.html.

Phillips, Jean. 2014. "Study Shows Tropical Cyclones Intensity Shifting Poleward." University of Wisconsin–Madison News. May 14, 2014. https://news.wisc.edu/study-shows-tropical-cyclone-intensity-shifting-poleward/.

RCraig09. 2018. "1851–2017 Atlantic Hurricanes and Tropical Storms by Month." Wikimedia Commons. September 7, 2018. https://commons.wikimedia.org/wiki/File:1851-2017_Atlantic_hurricanes_and_tropical_storms_by_month.png.

Schreiber, Melody. 2021. "Welcome to the Climate-Covid Convergence." *New Republic*, August 31, 2021. https://newrepublic.com/article/163474/hurricane-ida-climate-covid-wildfire.

Trimble, Megan. 2017. "Top 10 Deadliest Hurricanes in US History." *US News and World Report*, August 31, 2017. https://www.usnews.com/news/best-states/slideshows/the-deadliest-storms-in-us-history.

WDSU News. 2021. "Officials Say Grand Isle 'Uninhabitable,' 100% of Structures Damaged." Updated August 31, 2021. https://www.wdsu.com/article/grand-isle-uninhabitable-ida/37447820.

TORNADOES

Allaby, Michael. 2004. *Tornadoes*. Rev. ed. Facts on File Dangerous Weather Series, Facts on File Science Library, January 1, 2004. New York: Infobase.

CDC (Centers for Disease Control and Prevention). 2021. "Staying Safe in a Tornado." National Center for Environmental Health. Last reviewed March 29, 2021. https://www.cdc.gov/nceh/features/tornadosafety/index.html.

Centers, Josh. 2021. "How to Prepare for and Survive a Tornado." The Prepared, updated October 14, 2021. https://theprepared.com/emergencies/guides/tornado/.

Choi-Schagrin, Winston. 2022. "Are Tornadoes Changing along with Climate?" *New York Times*, March 23, 2022. https://www.nytimes.com/article/tornado-climate-change.html.

Edwards, Roger. 2021. "Tornado Safety." Storm Prediction Center, Norman, Oklahoma, NOAA (National Oceanic and Atmospheric Administration). Last updated March 21, 2021. https://www.spc.noaa.gov/faq/tornado/safety.html.

Elsner, J. B. 2021. "Is Climate Change Causing More Tornadoes?" Op-Ed. *New York Times*, Late Edition (East Coast), New York, NY. December 15, 2021. https://www.nytimes.com/2021/12/15/opinion/tornado-climate-change.html.

FEMA (Federal Emergency Management Agency). 2022. "Safe Room Publications and Resources." Last updated February 9, 2022. https://www.fema.gov/emergency-managers/risk-management/safe-rooms/resources.

Flavelle, Christopher. 2021. "We Know How to Protect People and Buildings against Tornadoes. So Why Don't We?" 2021 Climate Year in Review: Climate Forward, *New York Times*, December 29, 2021.

Kostigen, Thomas M. 2014. *Extreme Weather Survival Guide: Understand, Prepare, Survive, Recover*. Washington, DC: National Geographic.

Manzo, Daniel. 2021. "How to Prepare for and Survive a Tornado Outbreak." ABC News, March 25, 2021. https://abcnews.go.com/US/prepare-survive-tornado-outbreak/story?id=76629455.

Mersereau, Dennis. 2023. "Stuck in Your Car During a Tornado? Here's What You Should Do." Weather Network, June 3, 2023. Updated on June 20, 2023. https://www.theweathernetwork.com/en/news/weather/seasonal/stuck-in-your-car-during-a-tornado-heres-what-you-should-do.

National Weather Service. n.d.a. "Tornado Survivor Stories." Safety, National Weather Service, National Oceanic and Atmospheric Administration. Accessed December 22, 2021. https://www.weather.gov/safety/tornado-survivors.

National Weather Service. n.d.b. "What to Do During a Tornado." Safety, National Weather Service, National Oceanic and Atmospheric Administration. Accessed March 3, 2022. https://www.weather.gov/safety/tornado-during.

Norris, Anna. 2016. "What to Do If You See a Tornado While You're Driving." Tornado Safety and Preparedness, Weather Channel, February 25, 2016. https://weather.com/safety/tornado/news/what-to-do-see-tornado-while-driving.

HEAT

Atwoli, Lukoye, Abdullah H. Baqui, Thomas Benfield, Raffaella Bosurgi, Fiona Godlee, Stephen Hancocks, Richard Horton, et al. 2021. "Call for Emergency Action to Limit Global Temperature Increases, Restore Biodiversity, and Protect Health." *New England Journal of Medicine* 385: 1134–37. https://doi.org/10.1056/NEJMe2113200.

Berardelli, Jeff. 2021. "What Is a Heat Dome? Extreme Temperatures in the Pacific Northwest, Explained." CBS News, June 30, 2021. https://www.cbsnews.com/news/what-is-heat-dome-extreme-temperatures-pacific-northwest/.

Buis, Alan. 2022. "Too Hot to Handle: How Climate Change May Make Some Places Too Hot to Live." Ask NASA Climate, Jet Propulsion Laboratory, NASA, March 9, 2022. https://climate.nasa.gov/ask-nasa-climate/3151/too-hot-to-handle-how-climate-change-may-make-some-places-too-hot-to-live/.

Cara, Ed. 2021. "The Era of Climate Sickness Has Arrived." Gizmodo, October 20, 2021. https://gizmodo.com/the-era-of-climate-sickness-has-arrived-1847903223.

Cecco, Leyland. 2021. "'Heat Dome' Probably Killed 1bn Marine Animals on Canada Coast, Experts Say." *Guardian*, July 8, 2021. https://www.theguardian.com/environment/2021/jul/08/heat-dome-canada-pacific-northwest-animal-deaths.

EPA (US Environmental Protection Agency). 2017. "Technical Documentation: Heat-Related Deaths." https://www.epa.gov/sites/default/files/2017-01/documents/heat-deaths_documentation.pdf.

EPA (US Environmental Protection Agency). 2021. "Climate Change Indicators in the United States: Heat-Related Deaths." Climate Indicators. Updated April 21, 2021. https://www.epa.gov/climate-indicators/climate-change-indicators-heat-related-deaths.

Expatriate Group. 2019. "Expat Safety Tips: Heatwaves." Expatriate Healthcare. August 6, 2019. https://www.expatriatehealthcare.com/expat-safety-tips-heatwaves/.

Gorvett, Zarla. 2020. "The Troubling Ways a Heatwave Can Warp Your Mind." BBC Future, August 17, 2020. https://www.bbc.com/future/article/20200817-the-sinister-ways-heatwaves-warp-the-mind.

Huber, Matthew. 2022. "What Happens If the World Gets Too Hot for Animals to Survive?" *Mother Jones*, July 24, 2022. https://www.motherjones.com/politics/2022/07/what-happens-if-the-world-gets-too-hot-for-animals-to-survive/.

Irfan, Umair. 2021. "How Heat Waves Form, and How Climate Change Makes Them Worse." VOX, updated June 30, 2021. https://www.vox.com

/22538401/heat-wave-record-temperature-extreme-climate-change-drought.

Klinenberg, Eric. 2015. *Heat Wave: A Social Autopsy of Disaster in Chicago.* Chicago: University of Chicago Press.

Krajick, Kevin. 2020. "Potentially Fatal Combinations of Humidity and Heat Are Emerging across the Globe." State of the Planet, Columbia Climate School. May 8, 2020. https://news.climate.columbia.edu/2020/05/08/fatal-heat-humidity-emerging/.

Labbé, Stefan. 2021. "The B.C. Heat Dome Killed at Least 651,000 Farm Animals." *Squamish Chief*, September 28, 2021. https://www.squamishchief.com/highlights/the-bc-heat-dome-killed-at-least-651000-farm-animals-4467011.

The Lancet. n.d. "*The Lancet* Countdown on Health and Climate Change." Accessed November 17, 2021. https://www.thelancet.com/countdown-health-climate.

McCabe, Kirsty. 2021. "What Is a Heat Dome?" MetMatters, Royal Meterological Society, August 11, 2021. https://www.rmets.org/metmatters/what-heat-dome.

NOAA (National Oceanic and Atmospheric Administration). 2020. "Dangerous Humid Heat Extremes Occurring Decades before Expected." NOAA Research News. May 8, 2020. https://research.noaa.gov/article/ArtMID/587/ArticleID/2621/Dangerous-humid-heat-extremes-occurring-decades-before-expected.

NOAA (National Oceanic and Atmospheric Administration). 2021. "What Is a Heat Dome?" National Ocean Service. July 7, 2021. https://oceanservice.noaa.gov/facts/heat-dome.html.

Raymond, Colin, Tom Matthews, and Radley M. Horton. 2020. "The Emergence of Heat and Humidity Too Severe for Human Tolerance." *Science Advances* 6, no. 19. https://doi.org/10.1126/SCIADV.AAW1838.

Sengupta, Somini. 2018. "In India, Summer Heat May Soon Be Literally Unbearable." *New York Times*, July 17, 2018. https://www.nytimes.com/2018/07/17/climate/india-heat-wave-summer.html.

Tanenbaum, Michael. 2022. "Humans Can't Withstand Temperatures, Humidities as High as Once Thought, Penn State Study Finds." Philly Voice, March 4, 2022. https://www.phillyvoice.com/heat-stroke-climate-change-risk-high-temperature-humidity-penn-state/.

Wallace-Wells, David. 2020. *The Uninhabitable Earth: Life after Warming.* New York: Tim Duggan Books/Penguin Random House.

Waugh, Rob. 2021. "What Are 'Wet-Bulb Temperatures'—and Why Are They So Important for Climate Change?" Yahoo! News, July 19, 2021. https://news.yahoo.com/wet-bulb-temperature-162057698.html.

Weisgarber, Maria. 2021. "Scientists Warn Extreme Heat Wave That Preceded Lytton Fire May Not Be Isolated Event." CTV News Vancouver, July 1, 2021. https://bc.ctvnews.ca/scientists-warn-extreme-heat-wave-that-preceded-lytton-fire-may-not-be-isolated-event-1.5493667.

"Wet-Bulb Temperature—An Overview." n.d. ScienceDirect. Accessed November 2, 2021. https://www.sciencedirect.com/topics/engineering/wet-bulb-temperature.

Widerynski, Stasia, Paul Schramm, Kathryn Conlon, Rebecca Noe, Elena Grossman, Michelle Hawkins, Seema Nayak, Matthew Roach, and Asante Shipp Hilts. 2017. *The Use of Cooling Centers to Prevent Heat-Related Illness: Summary of Evidence and Strategies for Implementation.* Climate and Health Technical Report Series, Climate and Health Program, National Center for Environmental Health, Centers for Disease Control and Prevention, August 7, 2017. https://www.cdc.gov/climateandhealth/docs/UseOfCoolingCenters.pdf.

Williams, Augusta. 2019. "Heat Is a 'Silent Killer.'" T. H. Chan School of Public Health, Spotlight (podcast). January 24, 2019. https://www.hsph.harvard.edu/news/multimedia-article/extreme-heat-climate-change-health/.

Zhang, Yingxiao, and Allison L. Steiner. 2022. "Projected Climate-Driven Changes in Pollen Emission Season Length and Magnitude over the Continental United States." *Nature Communications* 13:1234. https://doi.org/10.1038/s41467-022-28764-0.

BAD AIR

California Air Resources Board (CARB). 2023. "Protecting Yourself from Wildfire Smoke." Office of Communications. Accessed August 24, 2023. https://ww2.arb.ca.gov/protecting-yourself-wildfire-smoke.

CDC (Centers for Disease Control and Prevention). 2020a. "Climate and Health: Air Pollution." National Center for Environmental Health. Last reviewed December 21, 2020. https://www.cdc.gov/climateandhealth/effects/air_pollution.htm.

CDC (Centers for Disease Control and Prevention). 2020b. "Fungal Diseases: Valley Fever (Coccidioidomycosis)." Division of Foodborne, Waterborne, and Environmental Diseases, National Center for Emerging and Zoonotic Infectious Diseases. Last reviewed December 29, 2020. https://www.cdc.gov/fungal/diseases/coccidioidomycosis/index.html.

EPA (US Environmental Protection Agency). 2018a. *Guide to Air Cleaners in the Home.* 2nd ed. Publication No. EPA-402-F-08-004. Office of Air and Radiation, Indoor Environments Division, May 2018. https://www.epa.gov/sites/default/files/2014-07/documents/aircleaners.pdf.

EPA (US Environmental Protection Agency). 2018b. "Wildfire Smoke Fact Sheet—Protecting Children from Wildfire Smoke and Ash." Publication No. EPA-452/F-18-006. AirNow. https://www.airnow.gov/publications/wildfire-guide-factsheets/protecting-children-from-wildfire-smoke-and-ash/.

EPA (US Environmental Protection Agency). 2019a. "Protect Your Pets from Wildfire Smoke." Publication No. EPA-452/F-19-002. AirNow. March 2019. https://www.airnow.gov/publications/wildfire-guide-factsheets/wildfire-smoke-protect-your-pets/.

EPA (US Environmental Protection Agency). 2019b. *Wildfire Smoke: A Guide for Public Health Officials.* Publication No. EPA-452/R-21-901. Research Triangle Park, NC: Office of Air Quality Planning and Standards, Health and Environmental Impacts Division. Last modified September 2021. https://www.airnow.gov/sites/default/files/2021-09/wildfire-smoke-guide_0.pdf.

EPA (US Environmental Protection Agency). 2021a. "Asthma Action Plan." Last updated March 24, 2021. https://www.epa.gov/asthma/asthma-action-plan.

EPA (US Environmental Protection Agency). 2021b. "Benefits and Costs of the Clean Air Act 1990–2020, the Second Prospective Study." Clean Air Act Overview. Last updated August 12, 2021. https://www.epa.gov/clean-air-act-overview/benefits-and-costs-clean-air-act-1990-2020-second-prospective-study.

EPA (US Environmental Protection Agency). 2021c. "2021 Changes to *Wildfire Smoke: A Guide for Public Health Officials.* AirNow. https://www.airnow.gov/sites/default/files/2021-09/wildfire-guide-information-corrections.pdf. Updates link: https://www.airnow.gov/wildfire-guide-post-publication-updates/.

Evans, Gareth. 2019. "A Toxic Crisis in America's Coal Country." BBC News, February 11, 2019. https://www.bbc.com/news/world-us-canada-47165522.

Lelieveld, J., K. Klingmüller, A. Pozzer, R. T. Burnett, A. Haines, and V. Ramanathan. 2019. "Effects of Fossil Fuel and Total Anthropogenic Emission Removal on Public Health and Climate." *Proceedings of the National Academy of Sciences* (PNAS) 116, no. 15: 7192–97. https://www.pnas.org/content/116/15/7192.

Mackenzie, Jillian, and Jeff Turrentine. 2021. "Air Pollution Facts: Everything You Need to Know." NRDC (National Resources Defense Council). June 22, 2021. https://www.nrdc.org/stories/air-pollution-everything-you-need-know.

Marot, Christelle. 2022. "Sandstorms, an Increasingly Common Global Disaster Exacerbated by Land Degradation." *Equal Times*, May 2, 2022. https://www.equaltimes.org/sandstorms-an-increasingly-common.

Mieszko the first. 2016. "File: Airborne-articulate-size-chart.jpg." Last edited on June 7, 2022. https://commons.wikimedia.org/wiki/File:Airborne-particulate-size-chart.jpg.

National Weather Service. n.d. "Why Air Quality Is Important." NOAA (National Oceanic and Atmospheric Administration). Accessed June 23, 2022. https://www.weather.gov/safety/airquality.

Roser, Max. 2021. "Data Review: How Many People Die from Air Pollution?" Our World in Data. November 25, 2021. https://ourworldindata.org/data-review-air-pollution-deaths.

Sommer, Lauren. 2022. "Rising Temperatures Prolong Pollen Season and Could Worsen Allergies." *All Things Considered*, NPR, updated April 14, 2022. https://www.npr.org/sections/health-shots/2022/03/15/1086733875/hotter-temps-bring-more-pollen-meaning-climate-change-will-intensify-allergy-sea.

UNEP (United Nations Environment Programme). 2019. "Air Pollution and Climate Change: Two Sides of the Same Coin." News and Stories. April 23, 2019. https://www.unep.org/news-and-stories/story/air-pollution-and-climate-change-two-sides-same-coin.

UNEP (United Nations Environment Programme). n.d. "Short-Lived Climate Pollutants (SLCPs)." Climate and Clean Air Coalition. Accessed November 1, 2021. https://www.ccacoalition.org/en/content/short-lived-climate-pollutants-slcps.

USDA (US Department of Agriculture). 2016. "Effects of Ozone Air Pollution on Plants." Plant Science Research, Agricultural Research Service, NC State University, Raleigh, NC. Last modified August 12, 2016. https://www.rst2.org/msu-ozone/university/ozonepdfs/1g27.pdf.

Wallace-Wells, David. 2020. *The Uninhabitable Earth: Life after Warming*. New York: Tim Duggan Books/Penguin Random House.

Yewell, John. 2020. "Climate Change Worsens Air Pollution, Extreme Weather, Experts Say." Environmental Factor, National Institute of Environmental Health Sciences. July 2020. https://factor.niehs.nih.gov/2020/7/science-highlights/climate-change/index.htm.

Fire

WILDFIRES

Associated Press. 2022. "As a Wildfire Closes In, New Mexico Residents Prepare to Flee." NPR, updated May 2, 2022. https://www.npr.org/2022/05/02/1096041597/new-mexico-wildfire.

Associated Press Television News. 2021. "Argentina Battles to Contain Patagonia Wildfires." RepublicWorld.com, last updated December 28, 2021. https://www.republicworld.com/world-news/south-america/argentina-battles-to-contain-patagonia-wildfires.html.

Atkinson, William. 2018. "The Link Between Power Lines and Wildfires." *Electrical Contractor Magazine*, November 25, 2018. https://www.ecmag.com/magazine/articles/article-detail/systems-link-between-power-lines-and-wildfires.

Cal Fire (California Department of Forestry and Fire Protection). 2023. "Remembering the Camp Fire." Our Impact, Cal Fire. Accessed August 24, 2023. https://www.fire.ca.gov/our-impact/remembering-the-camp-fire.

Cart, Julie. 2022. "Following Two Years after Australia's Lethal Black Summer Fires." *Discover*, April 9, 2022. https://www.discovermagazine.com/environment/following-two-years-after-australias-lethal-black-summer-fires.

Chappell, Bill. 2021. "Here's Why Firefighters Are Wrapping Sequoia Trees in Aluminum Blankets." Environment, NPR, September 20, 2021. https://www.npr.org/2021/09/20/1038972507/california-sequoia-trees-general-sherman-aluminum-blanket.

CNN. 2021. "In Pictures: Wildfires Raging in the West." Updated October 15, 2021. https://www.cnn.com/2021/07/19/us/gallery/western-wildfires-2021/index.html.

Coleman, Jude. 2022. "Australia's Epic Wildfires Expanded Ozone Hole and Cranked Up Global Heat." *Nature*, September 1, 2022. https://www.nature.com/articles/d41586-022-02782-w.

Daley, Jason. 2017. "Study Shows 84% of Wildfires Caused by Humans." *Smithsonian Magazine*, February 28, 2017. https://www.smithsonianmag.com/smart-news/study-shows-84-wildfires-caused-humans-180962315/.

DW News. 2021. "Summer Heat Waves Drive Wildfires in Europe." Deutsche Welle, August 1, 2021. https://www.dw.com/en/summer-heat-waves-drive-wildfires-in-europe/av-58720331.

EPA (US Environmental Protection Agency). 2021a. "Climate Change Indicators: Wildfires." Climate Indicators. Updated April 2021. https://www.epa.gov/climate-indicators/climate-change-indicators-wildfires.

EPA (US Environmental Protection Agency). 2021b. "EPA Wildfire Guide Factsheets." AirNow. Updated October 4, 2021. https://www.airnow.gov/wildfire-guide-factsheets/.

Expatriate Group. 2018. "Expat Safety Tips: Wildfires." Expatriate Health Care. August 14, 2018. https://www.expatriatehealthcare.com/expat-safety-tips-wildfires/.

Frontline Wildfire Defense. n.d. "What Causes Wildfires? Understanding Key Risk Factors." Wildfire News and Resources. Accessed June 23, 2022. https://www.frontlinewildfire.com/wildfire-news-and-resources/what-causes-wildfires/.

Ganey, Steve. 2019. "PG&E Lines Were Cause of Camp Fire That Killed 85 People, Cal Fire Investigation Finds." Associated Press via KTLA 5. Updated May 15, 2019. https://ktla.com/news/local-news/pge-lines-were-cause-of-camp-fire-that-killed-85-people-cal-fire-investigation-finds/.

Griffin, Jonathan. 2021. "New Timeline of Deadliest California Wildfire Could Guide Lifesaving Research and Action." NIST (National Institute of Standards and Technology). February 8, 2021; updated March 28, 2022. https://www.nist.gov/news-events/news/2021/02/new-timeline-deadliest-california-wildfire-could-guide-lifesaving-research.

Hardie, Alex, Amy Cassidy, and Hafsa Khalil. 2022. "Wildfires in the EU Have Nearly Quadrupled the 15-Year Average." CNN, July 27, 2022. https://www.cnn.com/2022/07/27/europe/wildfires-europe-increase-climate-intl/index.html.

Hawkins, Linnia R., John T. Abatzoglou, Sihan Li, and David E. Rupp. 2022. "Anthropogenic Influence on Recent Severe Autumn Fire Weather in the West Coast of the United States." *Geophysical Research Letters*, February 4, 2022. https://doi.org/10.1029/2021GL095496.

Henley, Jon. 2022. "Wildfires in Europe Burn Area Equivalent to One-Fifth of Belgium." *Guardian*, August 15, 2022. https://www.theguardian.com/world/2022/aug/15/wildfires-europe-burn-area-equivalent-one-fifth-belgium.

Irfan, Umair. 2020. "Australia's Hellish Heat Wave and Wildfires, Explained." Vox, updated January 6, 2020. https://www.vox.com/2019/12/30/21039298/40-celsius-australia-fires-2019-heatwave-climate-change.

Jones, Justice, April Saginor, and Brad Smith. 2011. "2011 Texas Wildfires: Common Denominators of Home Destruction." Texas A&M Forest Service. https://tfsweb.tamu.edu/uploadedFiles/TFSMain/Preparing_for_Wildfires/Prepare_Your_Home_for_Wildfires/Contact_Us/2011%20Texas%20Wildfires.pdf.

Kitzberger, Thomas. 2022. "Unprecedented Wildfires in Patagonia." Interview by Bobby Bascom. Living on Earth, week of January 28, 2022.

https://www.loe.org/shows/segments.html?programID=22-P13-00004&segmentID=1.
Mohler, Michael. 2019. "CAL FIRE Investigators Determine Cause of the Camp Fire." CAL FIRE news release, California Department of Forestry and Fire Protection, May 15, 2019. https://www.fire.ca.gov/media/5121/campfire_cause.pdf.
Mulhern, Owen. 2020. "Climate Change and the Australian Bushfires: A Visual Guide." Earth.org, July 27, 2020. https://earth.org/data_visualization/climate-change-and-the-australian-bushfires-a-visual-guide/.
National Interagency Fire Center. n.d. "National Fire News." Accessed July 18, 2022. https://www.nifc.gov/fire-information/nfn.
National Park Service. 2022. "Wildfire Causes and Evaluations." Last updated March 8, 2022. https://www.nps.gov/articles/wildfire-causes-and-evaluation.htm.
Orner, Eva, dir. 2021. *Burning*. Documentary film. 1 hour, 24 minutes. Culver City, CA: Amazon Studios. https://www.amazon.com/Burning-EVA-ORNER/dp/B09M45NHTX.
Porterfield, Carlie. 2022. "Heat Wave across Europe Sparks Wildfires and Heat-Related Deaths." *Forbes*, July 17, 2022. https://www.forbes.com/sites/carlieporterfield/2022/07/17/heat-wave-across-europe-sparks-wildfires-and-heat-related-deaths.
Ramsayer, Kate. 2017. "NASA Detects Drop in Global Fires." NASA's Goddard Space Flight Center, Greenbelt, MD. Last updated August 6, 2017. https://www.nasa.gov/feature/goddard/2017/nasa-detects-drop-in-global-fires.
Roberts, Joe. 2021. "The Most Dangerous States for Wildfires." MOVE.org. Accessed May 23, 2022. https://www.move.org/most-dangerous-states-wildfires/.
Sevrin, Anthony. 2022. "Why Are the 'Zombie' Fires in Siberia a Ticking Climate Bomb?" AXA Climate, May 30, 2022. https://www.climate.axa/articles/why-are-the-zombie-fires-in-siberia-a-ticking-climate-bomb.
Smith, Ed, Sonya Sistare, and Grant Nejedlo. 2018. *Fire Adapted Communities: The Next Step in Wildfire Preparedness, Elko County*. 2nd ed. Reno, NV: University of Nevada Cooperative Extension, March 2018. https://www.livingwithfire.com/wp-content/uploads/2018/10/Elko-FAC-sp1602.pdf.
UNEP (United Nations Environment Programme). 2022. *Spreading like Wildfire—the Rising Threat of Extraordinary Landscape Fires*. A UNEP Rapid Response Assessment. Nairobi: United Nations Envi-

ronment Programme. https://www.unep.org/resources/report/spreading-wildfire-rising-threat-extraordinary-landscape-fires.
Yewell, John. 2020. "Climate Change Worsens Air Pollution, Extreme Weather, Experts Say." Environmental Factor, National Institute of Environmental Health Sciences, July 2020. https://factor.niehs.nih.gov/2020/7/science-highlights/climate-change/index.htm.
Zamyatin, Yevgeny. 1967. "In Old Russia." In *The Dragon: Fifteen Stories*, 180. New York: Random House.

URBAN FIRESTORMS

Brasch, Sam, Joe Wertz, and Michael Elizabeth Sakas. 2022. "Air Quality, Wildfires and Building Codes: Here Are the Climate Bills Headed to Gov. Jared Polis's Desk." Colorado Public Radio News, May 12, 2022. https://www.cpr.org/2022/05/12/colorado-climate-bills-air-quality-wildfires-green-building-codes/.
"Common Causes of Electrical Fires." n.d. Morgan and Morgan. Accessed August 18, 2023. https://www.firelawsuit.com/latest-news/how-power-lines-can-spark-devastating-wildfires/.
Finley, Bruce. 2022. "'Spacing Here Was a Problem': Closely Built Homes Helped Marshall Firestorm Spread, Researchers Say." *Denver Post*, January 15, 2022; updated January 17, 2022. https://www.denverpost.com/2022/01/15/marshall-fire-insurance-losses-housing-density/.
Gabbert, Bill. 2022. "Marshall Fire Updated Damage Assessment: 1,084 Residences Destroyed." Wildfire Today, Wildfire News and Opinion. January 7, 2022. https://wildfiretoday.com/2022/01/07/marshall-fire-updated-damage-assessment-1084-residences-destroyed/.
McKinley, Carol, and Kyla Pearce. 2023. "Marshall Fire Had 2 Causes, Boulder County Sheriff's Office Says." *Denver Gazette*, June 21, 2023. https://denvergazette.com/news/wildfires/marshall-fire-cause-boulder-county-wildfire/article_d359f4a4-0585-11ee-8fa1-ab2162eb8260.html.
Wallace-Wells, David. 2022. "The Return of the Urban Firestorm: What Happened in Colorado Was Something Much Scarier Than a Wildfire." Intelligencer, Life after Warming, *New York Magazine*, January 1, 2022. https://nymag.com/intelligencer/2022/01/colorado-saw-the-return-of-the-urban-firestorm.html.

Water

SEA LEVEL RISE

Ariza, Mario Alejandro. 2020. "As Miami Keeps Building, Rising Seas Deepen Its Social Divide." Yale Environment 360. September 29, 2020.

https://e360.yale.edu/features/as-miami-keeps-building-rising-seas-deepen-its-social-divide.

Bender, Eric. 2021. "A $26-Billion Plan to Save the Houston Area from Rising Seas." Undark, June 14, 2021. https://undark.org/2021/06/14/texas-sized-effort-to-fend-off-rising-seas/.

Benn, Douglas I., Adrian Luckman, Jan A. Åström, Anna Crawford, Stephen L. Cornford, Suzanne L. Bevan, Rupert Gladstone, Thomas Zwinger, Karen Alley, Erin Pettit, and Jeremy Bassis. 2021. "Rapid Fragmentation of Thwaites Eastern Ice Shelf, West Antarctica." *Cryosphere* 16, no. 6: 2545–64. https://doi.org/10.5194/tc-16-2545-2022.

Boers, Niklas. 2021a. "Observation-Based Early-Warning Signals for a Collapse of the Atlantic Meridional Overturning Circulation." *Nature Climate Change* 11: 680–88. https://doi.org/10.1038/s41558-021-01097-4.

Boers, Niklas. 2021b. "Publisher Correction: Observation-Based Early-Warning Signals for a Collapse of the Atlantic Meridional Overturning Circulation." *Nature Climate Change* 11:1001. https://doi.org/10.1038/s41558-021-01184-6.

Climate Central. 2017. "These U.S. Cities Are Most Vulnerable to Major Coastal Flooding and Sea Level Rise." Research report. October 25, 2017. https://www.climatecentral.org/report/us-cities-most-vulnerable-major-coastal-flooding-sea-level-rise-21748.

Coastal Flood Resilience Project. 2022. *Proposed National Policies to Support Relocation of Communities as Sea Level Rises.* White paper. March 16, 2022. https://www.cfrp.info/_files/ugd/2450cf_1076f4e32d6d48d4ace774a20f403876.pdf.

Cornell University. 2017. "Rising Seas Could Result in 2 Billion Refugees by 2100." ScienceDaily, June 26, 2017. https://www.sciencedaily.com/releases/2017/06/170626105746.htm.

DeConto, Robert M., and David Pollard. 2016. "Contribution of Antarctica to Past and Future Sea-Level Rise." *Nature* 531:591–97. https://www.nature.com/articles/nature17145.

Delbert, Caroline. 2021. "The World's First True Floating Island City Could Reimagine Survival." *Popular Mechanics,* April 27, 2021. https://www.popularmechanics.com/science/environment/a36231724/maldives-floating-city/.

Dutch Docklands Maldives, and the Government of the Maldives. 2021. "Ministry of Tourism Permits Dutch Docklands Maldives to Run Floating City Development Based on the Integrated Tourism Model." Maldives Floating City. Press release, June 25, 2022. https://maldivesfloatingcity.com/wp-content/uploads/2021/03/20210314-Press-release-MFC-ENGLISH-N.pdf.

Englander, John. 2021. *Moving to Higher Ground: Rising Sea Level and the Path Forward*. Boca Raton, FL: Science Bookshelf.

Fitzsimmons, Tim. 2021. "Antarctic Ice Shelf Could Crack, Raise Seas by Feet within Decade, Scientists Warn." NBC News, December 15, 2021; updated December 16, 2021. http://www.nbcnews.com/science/environment/antarctic-ice-shelf.

Geisler, Charles, and Ben Currens. 2017. "Impediments to Inland Resettlement under Conditions of Accelerated Sea Level Rise." *Land Use Policy* 66:322–30. https://doi.org/10.1016/j.landusepol.2017.03.029.

Goodell, Jeff. 2017. *The Water Will Come: Rising Seas, Sinking Cities, and the Remaking of the Civilized World*. Boston: Little, Brown.

Hauer, Mathew E., Dean Hardy, Scott A. Kulp, Valerie Mueller, David J. Wrathall, and Peter U. Clark. 2021. "Assessing Population Exposure to Coastal Flooding Due to Sea Level Rise." *Nature Communications* 12:6900. https://www.nature.com/articles/s41467-021-27260-1.

Hesse, Hermann. 2008. *Siddhartha: An Indian Tale*. Translated by Gunther Olesch, Anke Dreher, Amy Coulter, Stefan Langer, and Semyon Chaichenets. Project Gutenberg. Updated January 23, 2013.

Kaplan, Sarah. 2021. "A Critical Atlantic Ocean System May Be Heading for Collapse from Climate Change, Study Finds." *Seattle Times*, August 5, 2021. https://www.seattletimes.com/nation-world/a-critical-atlantic-ocean-system-may-be-heading-for-collapse-from-climate-change-study-finds/.

Knowles, David. 2022. "Report: U.S. Will See 1 Foot of Sea Level Rise by 2050 because of Climate Change." Yahoo! News, February 15, 2022. https://www.yahoo.com/news/report-us-will-see-one-foot-of-sea-level-rise-by-2050-because-of-climate-change-195528451.html.

Kulp, Scott, Benjamin Strauss, Dyonishia Nieves, Shari Bell, and Dan Rizza. 2017. "These U.S. Cities Are Most Vulnerable to Major Coastal Flooding and Sea Level Rise." Climate Central. October 24, 2017. https://www.climatecentral.org/report/us-cities-most-vulnerable-major-coastal-flooding-sea-level-rise-21748.

Maldives Floating City. 2021. "Brain Coral: Concept Inspiration for Maldives Floating City." Accessed March 4, 2022. https://maldivesfloatingcity.com/.

Miami Beach Rising Above. 2017. "King Tides." Fact Sheet, April 11, 2017. https://www.miamibeachfl.gov/wp-content/uploads/2017/08/King-Tides-FactSheet-2-3.pdf.

Nicholls, R. J., S. Hanson, C. Herweijer, N. Patmore, S. Hallegatte, Jan Corfee-Morlot, Jean Chateau, and R. Muir-Wood. 2007. *Ranking of the World's Cities Most Exposed to Coastal Flooding Today and in the Fu-*

ture: Executive Summary. Extract from full report: OECD Environment Working Paper No. 1 (ENV/WKP(2007)1). Paris: Organisation for Economic Co-operation and Development (OECD). https://www.oecd.org/env/cc/39721444.pdf.

NOAA (National Oceanic and Atmospheric Administration). 2023. "What Is a King Tide?" Facts, National Ocean Service. Last updated June 1, 2023. https://oceanservice.noaa.gov/facts/kingtide.html.

Peters, Xander. 2021. "Rising Waters Set Stage for More Sea Walls in US Future." *Christian Science Monitor,* October 26, 2021. https://www.csmonitor.com/Environment/2021/1026/Rising-waters-set-stage-for-more-sea-walls-in-US-future.

Peterson, Jeff. 2022. "How to Help Communities Step Back from the Coast as Sea Levels Rise." *The Hill,* April 8, 2022. https://thehill.com/opinion/energy-environment/3262617-how-to-help-communities-step-back-from-the-coast-as-sea-levels-rise/.

Peterson, Jeffrey. 2019. *A New Coast: Strategies for Responding to Devastating Storms and Rising Seas.* Washington, DC: Island Press.

Pettit, Erin C., Christian Wild, Karen Alley, Atsuhiro Muto, Martin Truffer, Suzanne Louise Bevan, Jeremy N. Bassis, Anna Crawford, Ted A. Scambos, and Doug Benn (ITGC TARSAN Team). 2021. "C34A-07—Collapse of Thwaites Eastern Ice Shelf by Intersecting Fractures (Invited)." Paper presented at the American Geophysical Union Fall Meeting, New Orleans, Louisiana, December 15, 2021. https://agu.confex.com/agu/fm21/meetingapp.cgi/Paper/978762.

Pilkey, Orrin H., and Keith C. Pilkey. 2019. *Sea Level Rise: A Slow Tsunami on America's Shores.* Durham, NC: Duke University Press.

Pilkey, Orrin H., Linda Pilkey-Jarvis, and Keith C. Pilkey. 2016. *Retreat from a Rising Sea: Hard Choices in an Age of Climate Change.* New York: Columbia University Press.

Pilkey, Orrin H., and Robert Young. 2009. *The Rising Sea.* Washington, DC: Island Press.

Pilkey, Orrin H., Sr., Walter D. Pilkey, Orrin H. Pilkey Jr., and William J. Neal. 1984. *Coastal Design: A Guide for Planners, Developers, and Homeowners.* New York: Van Nostrand Reinhold.

Svokos, Alexandra. 2022. "Sanibel Island, Fort Myers Severely Damaged by Hurricane Ian in Lee County." ABC News, October 3, 2022. https://abcnews.go.com/US/sanibel-island-lee-county-facing-impacts-hurricane-ian/story?id=90698639.

Sweet, William V., Benjamin D. Hamlington, Robert E. Kopp, Christopher P. Weaver, Patrick L. Barnard, David Bekaert, William Brooks, et al. 2022. *Global and Regional Sea Level Rise Scenarios for the United*

States: Updated Mean Projections and Extreme Water Level Probabilities along U.S. Coastlines. NOAA Technical Report NOS 01. Silver Spring, MD: National Ocean Service, NOAA (National Oceanic and Atmospheric Administration). https://oceanservice.noaa.gov/hazards/sealevelrise/noaa-nos-techrpt01-global-regional-SLR-scenarios-US.pdf.

Tran, Viet Duc. 2013. "Which Coastal Cities Are at Highest Risk of Damaging Floods? New Study Crunches the Numbers." World Bank. August 19, 2013. https://www.worldbank.org/en/news/feature/2013/08/19/coastal-cities-at-highest-risk-floods.

USGS (US Geological Survey). 2018. "Coastal Landscape Response to Sea-Level Rise Assessment for the Northeastern United States." Woods Hole Coastal and Marine Science Center, December 30, 2018. Accessed November 19, 2021. https://www.usgs.gov/centers/whcmsc/science/coastal-landscape-response-sea-level-rise-assessment-northeastern-united.

Wanless, Harold R. 2021. "We Need the Courage to Truly Tackle Global Warming." The Invading Sea: Florida and the Climate Crisis. February 25, 2021. https://www.theinvadingsea.com/2021/02/25/we-need-the-courage-to-truly-tackle-global-warming/.

Wilson, Edward O. 2017. *Half-Earth: Our Planet's Fight for Life.* New York: Liveright/W. W. Norton.

OCEAN ACIDIFICATION

Dunagan, Christopher. 2019. "Rate of Ocean Acidification May Accelerate, Scientists Warn." *Salish Sea Currents Magazine, Encyclopedia of Puget Sound,* December 15, 2019. https://www.eopugetsound.org/magazine/IS/ocean-acdification.

Gattuso, Jean-Pierre, and Lina Hansson, eds. 2011. *Ocean Acidification.* New York: Oxford University Press.

Hawthorne, John. 2018. "Overfishing Is a Huge Problem. Here's What You Need to Know." *eco Magazine,* February 26, 2018. https://www.ecomagazine.com/news/policy/overfishing-is-a-huge-problem-here-s-what-you-need-to-know.

IPCC (Intergovernmental Panel on Climate Change). 2021a. *Climate Change 2021: The Physical Science Basis; Contribution of Working Group I to the Sixth Assessment Report of the Intergovernmental Panel on Climate Change,* edited by Valérie Masson-Delmotte, Panmao Zhai, Anna Pirani, Sarah L. Connors, Clotilde Péan, Sophie Berger, Nada Caud, et al. Cambridge: Cambridge University Press. https://www.ipcc.ch/report/ar6/wg1/#SPM.

Kennedy, Caitlyn. 2014. "Ocean Acidity Dissolving Tiny Snails' Protective Shell." Climate.gov, NOAA (National Oceanic and Atmospheric

Administration). May 8, 2014. https://www.climate.gov/news-features/featured-images/ocean-acidity-dissolving-tiny-snails%E2%80%99-protective-shell.

Klein, Alice. 2016. "Adding Rocks to Oceans Could De-acidify Water and Save Coral." *New Scientist*, May 10, 2016; amended May 11, 2016. https://www.newscientist.com/article/2087532-adding-rocks-to-oceans-could-de-acidify-water-and-save-coral/.

Knoll, Andrew H. 2021. *A Brief History of Earth: Four Billion Years in Eight Chapters*. New York: HarperCollins.

Mitloehner, Frank, Ermias Kebreab, and Michael Boccadoro. 2020. *Methane, Cows, and Climate Change: California Dairy's Path to Climate Neutrality*. CLEAR Center, University of California, Davis, September 2, 2020. https://clear.ucdavis.edu/sites/g/files/dgvnsk7876/files/inline-files/CLEAR-Center-Methane-Cows-Climate-Change-Sep-2-20_6.pdf.

Montserrat, Francesc, Phil Renforth, Jens Hartmann, Martine Leermakers, Pol Knops, and Filip J. R. Meysman. 2017. "Olivine Dissolution in Seawater: Implications for CO_2 Sequestration through Enhanced Weathering in Coastal Environments." *Environmental Science and Technology* 51, no. 7: 3960–72.

Nelson, Diane. 2021. "Feeding Cattle Seaweed Reduces Their Greenhouse Gas Emissions 82 Percent." College of Agricultural and Environmental Sciences, University of California, Davis. March 17, 2021. https://caes.ucdavis.edu/news/feeding-cattle-seaweed-reduces-their-greenhouse-gas-emissions-82-percent.

NOAA (National Oceanic and Atmospheric Administration). n.d. "Understanding Ocean Acidification." NOAA Fisheries. Accessed October 21, 2021. https://www.fisheries.noaa.gov/insight/understanding-ocean-acidification.

NOAA (National Oceanic and Atmospheric Administration. 2020. "Ocean Acidification." NOAA. Education. Last updated April 21, 2020. https://www.noaa.gov/education/resource-collections/ocean-coasts/ocean-acidification.

Ocean Portal Team. 2018. "Ocean Acidification." Smithsonian Ocean: Find Your Blue. April 2018. https://ocean.si.edu/ocean-life/invertebrates/ocean-acidification.

Waldbusser, George. 2013. "Study: Ocean Acidification Killing Oysters by Inhibiting Shell Formation." Oregon State University. June 11, 2013. https://today.oregonstate.edu/archives/2013/jun/study-ocean-acidification-killing-oysters-inhibiting-shell-formation-0

Woods Hole Oceanographic Institution. 2021. "Ocean Acidification." Know Your Ocean. Accessed July 2, 2023. https://www.whoi.edu/know

-your-ocean/ocean-topics/how-the-ocean-works/ocean-chemistry /ocean-acidification/.

MARINE HEAT WAVES

Bernton, Hal. 2022. "Alaska Cancels Bering Sea King and Snow Crab Seasons over Worries of Population Collapse." *Anchorage Daily News*, October 11, 2022.Updated October 16, 2022. https://www.adn.com /business-economy/2022/10/11/alaskas-bering-sea-king-and-snow-crab -seasons-canceled-over-population-collapse/.

Birkhead, Tim. 2020. "Worst Marine Heatwave on Record Killed One Million Seabirds in North Pacific Ocean." The Conversation, January 15, 2020. https://theconversation.com/worst-marine-heatwave-on-record -killed-one-million-seabirds-in-north-pacific-ocean-129842.

Bond, Nicholas. 2021. "The Blob: What Were Its Impacts on the Salish Sea?" In *State of the Salish Sea*, by K. L. Sobocinski, 134–35. Bellingham, WA: Salish Sea Institute, Western Washington University. http://doi .org/10.25710/vfhb-3a69.

Cornwall, Warren. 2019. "Ocean Heat Waves Like the Pacific's Deadly 'Blob' Could Become the New Normal." *Science*, January 31, 2019. https://www.science.org/content/article/ocean-heat-waves-pacific-s -deadly-blob-could-become-new-normal.

Derham, Kelly. 2021. "California's Disappearing Kelp Forests and Why It Matters for You." Sierra Club Pennsylvania. June 26, 2021. https://www .sierraclub.org/pennsylvania/blog/2021/06/california-s-disappearing -kelp-forests-and-why-it-matters-for-you.

Gentemann, Chelle L., Melanie R. Fewings, and Marisol García-Reyes. 2017. "Satellite Sea Surface Temperatures along the West Coast of the United States during the 2014–2016 Northeast Pacific Marine Heat Wave." *Geophysical Research Letters* 44:312–19. https://doi.org/10 .1002/2016GL071039.

Hobday, Alistair J., Lisa V. Alexander, Sarah E. Perkins, Dan A. Smale, Sandra C. Straub, Eric C. J. Oliver, Jessica A. Benthuysen, et al. 2016. "A Hierarchical Approach to Defining Marine Heatwaves." *Progress in Oceanography* 141:227–38. https://doi.org/10.1016/j.pocean.2015.12 .014.

IUCN (International Union for Conservation of Nature and Natural Resources). 2021. "Marine Heatwaves." IUCN Issues Briefs. Last reviewed October 2021. https://www.iucn.org/resources/issues-brief/marine -heatwaves.

Ma, Michelle. 2022. "New Study: 2021 Heat Wave Created 'Perfect Storm' for Shellfish Die-Off." Ecological Society of America. June 21, 2022.

https://www.esa.org/blog/2022/06/22/new-study-2021-heat-wave-created-perfect-storm-for-shellfish-die-off/.

Malmquist, David. 2022. "Increase in Marine Heat Waves Threatens Coastal Habitats." Phys.org. January 18, 2022. https://phys.org/news/2022-01-marine-threatens-coastal-habitats.html.

Marine Heatwaves International Working Group. n.d. "Marine Heatwaves Explained." Marine Heatwave News. Accessed January 30, 2022. http://www.marineheatwaves.org/all-about-mhws.html.

Mazzini, Piero L. F., and Cassia Pianca. 2022. "Marine Heatwaves in the Chesapeake Bay." *Frontiers in Marine Science* 8: 750265. https://doi.org/10.3389/fmars.2021.750265.

Naranjo, Laura. 2018. "The Blob." Earth Data, NASA. November 2, 2018. https://www.earthdata.nasa.gov/learn/sensing-our-planet/blob. https://www.fisheries.noaa.gov/feature-story/alaska-cod-populations-plummeted-during-blob-heatwave-new-study-aims-find-out-why.

NOAA (National Oceanic and Atmospheric Administration). 2019. "So What Are Marine Heat Waves? A noaa Scientist Explains." NOAA Research News. October 8, 2019. https://research.noaa.gov/article/ArtMID/587/ArticleID/2559/So-what-are-marine-heat-waves.

NOAA (National Oceanic and Atmospheric Administration). 2021. "Why Should We Care about the Ocean?" Ocean Facts, Our World Ocean, National Ocean Service. February 26, 2021. https://oceanservice.noaa.gov/facts/why-care-about-ocean.html.

NOAA Fisheries. 2019. "Alaska Cod Populations Plummeted During The Blob Heatwave—New Study Aims to Find Out Why." NOAA Fisheries, November 8, 2019. Last updated by Alaska Fisheries Science Center, December 7, 2022. Oliver, Eric C. J., Markus G. Donat, Michael T. Burrows, Pippa J. Moore, Dan A. Smale, Lisa V. Alexander, Jessica A. Benthuysen, et al. 2018. "Longer and More Frequent Marine Heatwaves over the Past Century." *Nature Communications* 9:1324. https://doi.org/10.1038/s41467-018-03732-9.

Popovich, Nadja, and Winston Choi-Schagrin. 2021. "Hidden Toll of the Northwest Heat Wave: Hundreds of Extra Deaths." *New York Times*, August 11, 2021. https://www.nytimes.com/interactive/2021/08/11/climate/deaths-pacific-northwest-heat-wave.html.

Spillman, Claire M., Grant A. Smith, Alistair J. Hobday, and Jason R. Hartog. 2021. "Onset and Decline Rates of Marine Heatwave: Global Trends, Seasonal Forecasts and Marine Management." *Frontiers in Climate* 3 (December 24, 2021). https://doi.org/10.3389/fclim.2021.801217.

Turner, Nicholas. 2022. "Ocean Heat Waves Trigger 'Squid Bloom' along Pacific Northwest Coast, Scientists Say." *Seattle Times*, February 3,

2022. https://www.seattletimes.com/seattle-news/environment/ocean-heat-waves-trigger-squid-bloom-along-west-coast-scientists-say/.
UN (United Nations). n.d. "How Is Climate Change Impacting the World's Ocean?" Climate Action. Accessed July 13, 2022. https://www.un.org/en/climatechange/science/climate-issues/ocean-impacts.
Xiang, Chloe. 2022. "Alaska Canceled Snow Crab Season for First Time Ever Because All the Crabs Are Gone." Motherboard, Vice, October 14, 2022. https://www.vice.com/en/article/7k87gy/alaska-canceled-snow-crab-season-for-the-first-time-ever-because-all-the-crabs-are-gone.

TSUNAMIS

Alaska Department of Natural Resources. 2022. "Barry Arm Landslide and Tsunami Hazard." Geological and Geophysical Surveys. Updated June 3, 2022. https://dggs.alaska.gov/hazards/barry-arm-landslide.html.
Brannen, Peter. 2017. *The Ends of the World: Volcanic Apocalypses, Lethal Oceans, and Our Quest to Understand Earth's Past Mass Extinctions.* New York: Ecco/HarperCollins.
Briscoe, Tony. 2019. "Lake Michigan's Deadly 'Freak Wave' of 1954 Is Chicago's Folklore. Turns Out It Was a Meteotsunami. And They Happen Pretty Often." *Chicago Tribune*, April 25, 2019. https://www.chicagotribune.com/news/ct-met-lake-michigan-meteotsunami-waves-20190415-story.html.
Brocher, Thomas M., John R. Filson, Gary S. Fuis, Peter J. Haeussler, Thomas L. Holzer, George Plafker, and J. Luke Blair. 2014. *The 1964 Great Alaska Earthquake and Tsunamis—A Modern Perspective and Enduring Legacies.* United States Geological Survey Fact Sheet 2014–3018. Reston, VA: US Geological Survey. https://dx.doi.org/10.3133/fs20143018.
Dusek, Gregory, Christopher DiVeglio, Louis Licate, Lorraine Heilman, Katie Kirk, Christopher Paternostro, and Ashley Miller. 2019. "A Meteotsunami Climatology along the U. S. East Coast." *Bulletin of the American Meteorological Society* 100, 7: 1329–45. https://doi.org/10.1175/BAMS-D-18-0206.1.
Expatriate Group. 2018. "Expat Safety Tips: What to Do During a Tsunami." Expatriate Healthcare. April 13, 2018. https://www.expatriatehealthcare.com/expat-safety-tips-what-to-do-during-a-tsunami/.
Higman, Bretwood, Marten Geertsema, Dan Shugar, Patrick Lynett, and Anja Dufresne. 2019. "The 2015 Taan Fiord Landslide and Tsunami." *Alaska Park Science* 18, no. 1: 6–15. https://www.nps.gov/articles/aps-18-1-2.htm.
Irwin, Aisling. 2021. "Why Climate Change Could Make Mediterranean Atmospheric 'Meteotsunamis' More Common." *Horizon: The EU Re-*

search and Innovation Magazine, May 11, 2021. https://ec.europa.eu/research-and-innovation/en/horizon-magazine/why-climate-change-could-make-mediterranean-atmospheric-meteotsunamis-more-common.

McGuire, Bill. 2012. *Waking the Giant: How a Changing Climate Triggers Earthquakes, Tsunamis, and Volcanoes*. Oxford: Oxford University Press.

Miller, Don J. 1960. *Giant Waves in Lituya Bay, Alaska*. United States Geological Survey Professional Paper 354-C. Washington, DC: US Government Printing Office. https://doi.org/10.3133/pp354C.

Mulligan, Ryan P., and Andy Take. 2020. "The World's Biggest Waves: How Climate Change Could Trigger Large Landslides and 'Mega-Tsunamis.'" The Conversation, August 17, 2020. https://theconversation.com/the-worlds-biggest-waves-how-climate-change-could-trigger-large-landslides-and-mega-tsunamis-115882.

NOAA (National Oceanic and Atmospheric Administration). 2015. "What Is a Meteotsunami?" National Tsunami Hazard Mitigation Program, National Weather Service, August 2015. https://nws.weather.gov/nthmp/documents/meteotsunamis.pdf.

NOAA (National Oceanic and Atmospheric Administration). 2016. "Tsunami Historical Series: Sumatra—2004." Science on a Sphere, November 23, 2016. Accessed June 14, 2022. https://sos.noaa.gov/catalog/datasets/tsunami-historical-series-sumatra-2004/.

NOAA (National Oceanic and Atmospheric Administration). 2021. "New Study Shows Promise of Forecasting Meteotsunamis: Photographer's Images Aid Research on Rare Great Lakes Wave." NOAA Research News. March 31, 2021. https://research.noaa.gov/article/ArtMID/587/ArticleID/2738/NOAA-research-shows-promise-of-forecasting-weather-driven-tsunamis.

Parry, Richard Lloyd. 2017. Prologue to *Ghosts of the Tsunami: Death and Life in Japan's Disaster Zone*. New York: MCD/Farrar, Straus and Giroux.

Petersen, Victoria. 2020. "Climate Change Intensifies Tsunami Threat in Alaska: As Glaciers Retreat and Permafrost Thaws, Massive Landslides Threaten Coastal Communities." *High Country News*, November 9, 2020. https://www.hcn.org/issues/52.12/north-climate-change-intensifies-tsunami-threat-in-alaska.

Pilarczyk, Jessica E., Yuki Sawai, Yuichi Namegaya, Toru Tamura, Koichiro Tanigawa, Dan Matsumoto, Tetsuya Shinozaki, et al. 2021. "A Further Source of Tokyo Earthquakes and Pacific Ocean Tsunamis." *Nature Geoscience* 14:796–800. https://doi.org/10.1038/s41561-021-00812-2.

Šepić, Jadranka, and Mirko Orlić. n.d. "Meteorological Tsunamis in the Adriatic Sea: Catalogue of Meteorological Tsunamis in Croatian Coastal

Waters." Institute of Oceanography and Fisheries Split, and Department of Geophysics, Faculty of Science, University of Zagreb. http://jadran.izor.hr/~sepic/meteotsunami_catalogue/#.

USGS (US Geological Survey). n.d. "M9.2 Alaska Earthquake and Tsunami of March 27, 1964." Earthquake Hazards Program. Accessed February 4, 2022. https://earthquake.usgs.gov/earthquakes/events/alaska1964/.

FLOODS

ABC News. 2021. "Timelapse Video of Catastrophic Flooding in Waverly TN on Aug 21." August 21, 2021. YouTube video, uploaded September 2, 2021. https://www.youtube.com/watch?v=RoZFKtNT34I.

Atwood, Margaret. 2006. *The Penelopiad*. Edinburgh: Canongate U.S.

Bacon, John, and Yue Stella Yu. 2021. "Floodwaters in Waverly, Tennessee, Shatter Homes and Families; Dozens Missing: 'We Will Persevere.'" *USA Today*, August 23, 2021. https://www.usatoday.com/story/news/nation/2021/08/23/waverly-tennessee-flooding-search-and-rescue-continues/8240572002/.

Bovarnick, Ben, Shiva Polefka, and Arpita Bhattacharyya. 2014. *Rising Waters, Rising Threat: How Climate Change Endangers America's Neglected Wastewater Infrastructure*. Washington, DC: Center for American Progress, October 2014. https://www.americanprogress.org/wp-content/uploads/sites/2/2014/10/wastewater-report.pdf.

Burke, Jason. 2022. "After the Relentless Rain, South Africa Sounds the Alarm on the Climate Crisis." *Guardian*, April 24, 2022. https://www.theguardian.com/world/2022/apr/24/south-africa-floods-rain-climate-crisis-extreme-weather.

Cappucci, Matthew. 2023. "California Sees Major Drought Improvement as Onslaught of Storms Ends." *Washington Post*, January 19, 2023. https://www.washingtonpost.com/weather/2023/01/19/california-flooding-drought-atmospheric-rivers/.

Coury, Nic, and Stefanie Dazio. 2023. "Photos: Atmospheric River Leaves California Inundated, with Another in the Forecast." PBS News Hour, March 12, 2012. https://www.pbs.org/newshour/nation/photos-atmospheric-river-leaves-california-inundated-with-another-in-the-forecast.

Dance, Scott. 2023. "Recently Parched Mississippi River Faces Major Floods as Record Snows Melt." *Washington Post*, April 27, 2023. https://www.washingtonpost.com/weather/2023/04/27/mississippi-river-flooding-minnesota-wisconsin-iowa/.

Dennis, Brady. 2022. "'They Are Not Slowing Down': The Rise of Billion-Dollar Disasters." *Washington Post*, August 4, 2022. https://www

.washingtonpost.com/climate-environment/interactive/2022/billion-dollar-disasters/.
Environment Florida Research and Policy Center. 2017. "Hurricane Irma and Sewage Spills." September 20, 2017. Accessed June 14, 2022. https://environmentamerica.org/florida/center/resources/hurricane-irma-and-sewage-spills/.
Environment Florida, Florida PIRG, and Frontier Group. 2017. "Hurricane Irma and Sewage Spills." Fact Sheet, September 21, 2017. Accessed July 2, 2023. https://publicinterestnetwork.org/wp-content/uploads/2017/09/Irma-sewage-factsheet-9.21-Environment-Florida-Final.pdf.
Expatriate Group. 2018. "Expat Safety Tips: What to Do During a Flood." Expatriate Healthcare. May 10, 2018. https://www.expatriatehealthcare.com/expat-safety-tips-what-to-do-during-a-flood/.
FEMA Preparedness Community. n.d. "Flood | Vehicle (Do Not Drive in Floodwaters; 'Turn Around, Don't Drown!')." Protective Actions, Federal Emergency Management Agency (FEMA). Accessed June 13, 2022. https://community.fema.gov/ProtectiveActions/s/article/Flood-Vehicle-Do-Not-Drive-in-Floodwaters-Turn-Around-Don-t-Drown.
First Street Foundation. 2021. *The 3rd National Risk Assessment: Infrastructure on the Brink*. October 2021. https://assets.firststreet.org/uploads/2021/09/The-3rd-National-Risk-Assessment-Infrastructure-on-the-Brink.pdf.
First Street Foundation. n.d. "What Will Climate Change Cost You?" Flood Factor. Accessed July 2, 2023. https://riskfactor.com/?utm_source=floodfactor.
Harvey, Chelsea. 2023. "15 Million People Are at Risk from Bursting Glacial Lakes." *Scientific American*, February 8, 2023. https://www.scientificamerican.com/article/15-million-people-are-at-risk-from-bursting-glacial-lakes/.
Labbé, Stefan. 2021. "The B.C. Heat Dome Killed at Least 651,000 Farm Animals." *Squamish Chief*, September 28, 2021. https://www.squamishchief.com/highlights/the-bc-heat-dome-killed-at-least-651000-farm-animals-4467011.
Masters, Jeff. 2021. "Extreme Rainfall in China: Over 25 Inches Falls in 24 Hours, Leaving 33 Dead." Yale Climate Connections. July 22, 2021. https://yaleclimateconnections.org/2021/07/extreme-rainfall-in-china-over-25-inches-falls-in-24-hours-leaving-33-dead/.
Miller, Brandon, Judson Jones, Sophia Saifi, and Kathleen Magramo. 2022. "Pakistan's Deadly Floods Have Created a Massive 100-km-Wide Inland Lake, Satellite Images Show." CNN, August 31, 2022. https://www.cnn.com/2022/08/31/asia/pakistan-floods-forms-inland-lake-satellite-intl-hnk.

NOAA (National Oceanic and Atmospheric Administration). n.d. "Severe Weather 101: Flood Types." NOAA National Severe Storms Laboratory. Accessed June 13, 2022. https://www.nssl.noaa.gov/education/svrwx101/floods/types/.

OCHA (United Nations Office for the Coordination of Humanitarian Affairs). 2022. "Pakistan Monsoon Floods 2022 Islamic Relief Pakistan (12 October, 2022)." Situation report. ReliefWeb. October 13, 2022. https://reliefweb.int/report/pakistan/pakistan-monsoon-floods-2022-islamic-relief-pakistan-12-october-2022.

Pratt, Sara E. 2022. "Devastating Floods in Pakistan." Earth Observatory, NASA (National Aeronautics and Space Administration). August 4–31, 2022. https://earthobservatory.nasa.gov/images/150279/devastating-floods-in-pakist.

Ramirez, Rachel, and Angela Dewan. 2022. "Pakistan Emits Less Than 1% of the World's Planet-Warming Gases. It's Now Drowning." CNN, August 31, 2022. https://www.cnn.com/2022/08/30/asia/pakistan-climate-crisis-floods-justice-intl/index.html.

Smith, Charlie. 2021. "Climate Stability Becomes a Relic of the Past in B.C. Due to Loss of Hydrologic Stationarity." *Georgia Straight*, Vancouver Free Press, December 1, 2021. https://www.straight.com/news/climate-stability-becomes-a-relic-of-past-in-bc-due-to-loss-of-hydrologic-stationarity.

Touma, Danielle, Samantha Stevenson, Daniel L. Swain, Deepti Singh, Dmitri A. Kalashnikov, and Xingying Huang. 2022. "Climate Change Increases Risk of Extreme Rainfall Following Wildfire in the Western United States." *Science Advances* 8, no. 13. https://www.science.org/doi/10.1126/sciadv.abm0320.

USGS (US Geological Survey). 2021. "Atmospheric River Rating System Chart." Communications and Publishing. December 13, 2021. https://www.usgs.gov/media/images/atmospheric-river-rating-system-chart.

Weather Channel. 2022. "New York City Will Distribute Inflatable Dams to Combat Flooding Damage." Video. July 22, 2022. https://weather.com/news/weather/video/new-york-city-will-distribute-inflatable-dams-to-combat-flooding-damage.

Weill, Alexandra (Allie). 2021. "Rivers in the Sky: 6 Facts You Should Know about Atmospheric Rivers." Communications and Publishing, US Geological Survey. December 14, 2021. https://www.usgs.gov/news/featured-story/rivers-sky-6-facts-you-should-know-about-atmospheric-rivers.

DROUGHT

Active Wild Admin. 2020. "North American Deserts: Facts on the Four Major Deserts of North America." Active Wild. September 25, 2020. https://www.activewild.com/north-american-deserts/.

Amadeo, Kimberly. 2020. "Drought's Effect on the Economy and You." The Balance. Updated November 25, 2020. https://www.thebalance.com/drought-definition-effects-examples-and-solutions-4157896.

Averyt, Kristen, Jeremy Fisher, Annette Huber-Lee, Aurana Lewis, Jordan Macknick, Nadia Madden, John Rogers, and Stacy Tellinghuisen. 2011. *Freshwater Use by U. S. Power Plants: Electricity's Thirst for a Precious Resource.* A Report of the Energy and Water in a Warming World Initiative, November 2011. Cambridge, MA: Union of Concerned Scientists. https://www3.epa.gov/region1/npdes/merrimackstation/pdfs/ar/AR-1501.pdf.

Baker, Aryn, and Mbar Roubab. 2019. "Can a 4,815-Mile Wall of Trees Help Curb Climate Change in Africa?" 2050: The Fight for Earth, *Time*, September 12, 2019. https://time.com/5669033/great-green-wall-africa/.

Budryk, Zack. 2022. "DRIED UP: Lakes Mead and Powell Are at the Epicenter of the Biggest Western Drought in History." *The Hill*, August 11, 2022. https://thehill.com/policy/energy-environment/3587785-dried-up-lakes-mead-and-powell-are-at-the-epicenter-of-the-biggest-western-drought-in-history/News%20Alerts/.

Center for Climate and Energy Solutions (C2ES). n.d. "Drought and Climate Change." Accessed January 11, 2022. https://www.c2es.org/content/drought-and-climate-change/.

DroughtScape. Winter 2022. Quarterly Newsletter, National Drought Mitigation Center, University of Nebraska-Lincoln. Accessed January 20, 2022. https://drought.unl.edu/Publications/DroughtScape.aspx.

Editors of Encyclopaedia Britannica. n.d. "List of Deserts." *Britannica.* Accessed January 11, 2022. https://www.britannica.com/topic/list-of-deserts-1854209.

EPA (US Environmental Protection Agency). 2021. "Climate Change Indicators: U.S. and Global Precipitation." Last updated July 17, 2021. https://www.epa.gov/climate-indicators/climate-change-indicators-us-and-global-precipitation.

Expatriate Group. 2019. "Expat Safety Tips: Drought." Expatriate Healthcare. January 30, 2019. https://www.expatriatehealthcare.com/expat-safety-tips-drought/.

Hill, Jessica. 2022. "Officials Explain 'Dead Pool' and How to Stop It in Lake Mead." *Las Vegas Sun*, May 26, 2022. https://lasvegassun.com/news/2022/may/26/understanding-dead-pool-and-how-water-officials-ar/.

IPCC (Intergovernmental Panel on Climate Change). 2021b. "Summary for Policymakers." In *Climate Change 2021: The Physical Science Basis; Contribution of Working Group I to the Sixth Assessment Report of the Intergovernmental Panel on Climate Change*, edited by Valérie Masson-Delmotte, Panmao Zhai, Anna Pirani, Sarah L. Connors, Clotilde Péan, Sophie Berger, Nada Caud, et al., 3–32. Cambridge: Cambridge University Press. https://www.ipcc.ch/report/sixth-assessment-report-working-group-i/. htpps://doi.org/10.1017/9781009157896.001.

Kiprop, Victor. 2018. "The Most Drought Prone Countries in the World." WorldAtlas. March 23, 2018. https://www.worldatlas.com/articles/the-most-drought-prone-countries-in-the-world.html.

Massachusetts Emergency Management Agency. n.d. "Drought Safety Tips." Mass.gov. Accessed December 12, 2021. https://www.mass.gov/info-details/drought-safety-tips.

Mukherjee, Sourav, and Ashok Kumar Mishra. 2022. "A Multivariate Flash Drought Indicator for Identifying Global Hotspots and Associated Climate Controls." *Geophysical Research Letters* 49, no. 2: E2021GL096804. https://doi.org/10.1029/2021GL096804.

National Weather Service. n.d. "Drought Types." Accessed February 7, 2022. https://www.weather.gov/safety/drought-types.

NPS (National Park Service). 2020. "Overview of Lake Mead." Last updated September 2, 2020. https://www.nps.gov/lake/learn/nature/overview-of-lake-mead.htm.

Tinker, Richard. 2022. "U.S. Drought Monitor." National Drought Mitigation Center, University of Nebraska-Lincoln. Map released January 6, 2022. https://droughtmonitor.unl.edu/.

UNICEF (The United Nations Children's Fund). 2022. "Prolonged Drought Pushing Families in Ethiopia to the Brink." Press Release, February 1, 2022. https://www.unicef.org/press-releases/prolonged-drought-pushing-families-ethiopia-brink.

US Department of Homeland Security. "Drought." n.d. Ready.gov. Last updated May 5, 2022. https://www.ready.gov/drought.

USGS (US Geological Survey). n.d. "What Causes Drought?" Frequently Asked Questions. Accessed January 10, 2022. https://www.usgs.gov/faqs/what-causes-drought.

Washington State Government. n.d. "Drought." Washington Military Department, Emergency Management Division. Accessed August 24, 2023. https://mil.wa.gov/drought#:~:text=Unlike%20most%20states%2C%20Washington%20has,below%2075%20percent%20of%20normal.

WBTF Media. n.d. "Deserts of the World." The 7 Continents of the World. Accessed January 11, 2022. https://www.whatarethe7continents.com/deserts-of-the-world/.

Woloszyn, Molly, Jesse E. Bell, Armanda Sheffield, Rachel Lookadoo, and Keith Hansen. 2022. "Advancing Drought Science and Preparedness Across the Nation: Focus on Public Health." Paper presented at the American Geophysical Union (AGU) Fall Meeting 2022, Chicago, IL, December 12–16, 2022 (id.SY16A-03). https://ui.adsabs.harvard.edu/abs/2022AGUFMSY16A..03W/abstract.

WATER SUPPLY

Breitberg, Denise, Lisa A. Levin, Andreas Oschlies, Marilaure Grégoire, Francisco P. Chavez, Daniel J. Conley, Véronique Garçon, et al. 2018. "Declining Oxygen in the Global Ocean and Coastal Waters." *Science* 359 (6371): eaam7240. https://doi.org/10.1126/science.aam7240.

Brown, Thomas C., Romano Foti, and Jorge A. Ramirez. 2013. "Projected Freshwater Withdrawals in the United States under a Changing Climate." *Water Resources Research* 49, no. 3: 1259–76. https://doi.org/10.1002/wrcr.20076.

CDC (Centers for Disease Control and Prevention). n.d. "Water Use Around the World." Global Health. Last reviewed January 2, 2020. https://www.cdc.gov/globalhealth/infographics/food-water/water_use.htm.

Chislock, Michael F., Enrique Dester, Rachel A. Zitomer, and Alan E. Wilson. 2013. "Eutrophication: Causes, Consequences, and Controls in Aquatic Ecosystems." *Nature Education Knowledge* 4, no. 4: 10. https://www.nature.com/scitable/knowledge/library/eutrophication-causes-consequences-and-controls-in-aquatic-102364466/.

Chow, Alex T.-S., Tanju Karanfil, and Randy A. Dahlgren. 2021. "Wildfires Are Threatening Municipal Water Supplies." *Eos*, August 12, 2021. https://doi.org/10.1029/2021EO161894.

Clark, Larry. 2021. "The Great Recession." *Washington State Magazine*, Winter 2021. https://magazine.wsu.edu/2021/11/08/the-great-recession/.

Dimick, Dennis. 2015. "Lack of Snow Leaves California's 'Water Tower' Running Low." *National Geographic*, March 4, 2015. https://www.nationalgeographic.com/science/article/150304-snow-snowpack-california-drought-groundwater-crisis.

Dixit, Kunda. 2019. "Himalayan Glaciers on Pace for Catastrophic Meltdown This Century, Report Warns." Inside Climate News, February 5, 2019. https://insideclimatenews.org/news/05022019/everest-himalayan

-mountains-climate-change-glacier-ice-loss-water-crisis-icimod-global-warming-report/.
Doshi, Rush, Alexis Dale-Huang, and Gaoqi Zhang. 2021. *Northern Expedition—China's Arctic Activities and Ambitions*. Foreign Policy at Brookings, the Brookings Institution, April 2021. https://www.brookings.edu/wp-content/uploads/2021/04/FP_20210412_china_arctic.pdf.
Engineering for Change. n.d. "Warka Water Tower." Solutions Library. Accessed April 20, 2022. https://www.engineeringforchange.org/solutions/product/warka-water-tower.
EPA (US Environmental Protection Agency). 2018. *U.S. Action Plan for Lake Erie: Commitments and Strategy for Phosphorus Reduction*. Great Lakes National Program Office, February 2018. https://www.epa.gov/sites/default/files/2018-03/documents/us_dap_final_march_1.pdf.
EPA (US Environmental Protection Agency). 2021. "Climate Impacts on Water Utilities." EPA Climate Change Adaptation Resource Center (ARC-X). Last updated August 31, 2021. https://www.epa.gov/arc-x/climate-impacts-water-utilities.
Fecht, Sarah. 2019. "How Climate Change Impacts Our Water." State of the Planet, Columbia Climate School. September 23, 2019. https://news.climate.columbia.edu/2019/09/23/climate-change-impacts-water/.
Gies, Erica. 2018. "Like Oceans, Freshwater Is Also Acidifying." *Scientific American*, January 11, 2018. https://www.scientificamerican.com/article/like-oceans-freshwater-is-also-acidifying/.
Kolbert, Elizabeth. 2019. "The Ice Stupas, Artificial Glaciers at the Edge of the Himalayas." *New Yorker*, May 13, 2019. https://www.newyorker.com/magazine/2019/05/20/the-art-of-building-artificial-glaciers.
Lall, U., T. Johnson, P. Colohan, A. Aghakouchak, C. Brown, G. McCabe, R. Pulwarty, and A. Sankarasubramanian. 2018. "Water." In *Impacts, Risks, and Adaptation in the United States: Fourth National Climate Assessment*, Vol. 2, edited by D. R. Reidmiller, C. W. Avery, D. R. Easterling, K. E. Kunkel, K. L. M. Lewis, T. K. Maycock, and B. C. Stewart, 145–73. U.S. Washington, DC: Global Change Research Program. https://doi.org/10.7930/NCA4.2018.CH3.
Miller, Elizabeth. 2021. "Climate Change Is Acidifying and Contaminating Drinking Water and Alpine Ecosystems." *Scientific American*, November 4, 2021. https://www.scientificamerican.com/article/climate-change-is-acidifying-and-contaminating-drinking-water-and-alpine-ecosystems/.
The Nature Conservancy. n.d. "Gulf of Mexico Dead Zone." Accessed March 15, 2022. https://www.nature.org/en-us/about-us/where-we-work/priority-landscapes/gulf-of-mexico/stories-in-the-gulf-of-mexico/gulf-of-mexico-dead-zone/.

NOAA (National Oceanic and Atmospheric Administration). 2021. "Larger-Than-Average Gulf of Mexico Dead Zone Measured." NOAA news release. August 3, 2021. https://www.noaa.gov/news-release/larger-than-average-gulf-of-mexico-dead-zone-measured.
NSF (National Science Foundation). 2021. "World's Lakes Losing Oxygen as Planet Warms." Research News, Division of Environmental Biology. June 17, 2021. https://beta.nsf.gov/news/worlds-lakes-losing-oxygen-planet-warms.
Oschlies, Andreas, Peter Brandt, Lothar Stramma, and Sunke Schmidtko. 2018. "Guest Post: How Global Warming Is Causing Ocean Oxygen Levels to Fall." Carbon Brief. June 15, 2018. https://www.carbonbrief.org/guest-post-how-global-warming-is-causing-ocean-oxygen-levels-to-fall/.
Pleitgen, Frederik, Claudia Otto, Angela Dewan, and Mohammed Tawfeeq. 2021. "The Middle East Is Running out of Water, and Parts of It Are Becoming Uninhabitable." CNN, updated August 22, 2021. https://www.cnn.com/2021/08/22/middleeast/middle-east-climate-water-shortage-iran-urmia-intl/index.html.
Rowe, Mark D., Reagan M. Errera, Edward S. Rutherford, Ashley K. Elgin, Darren J. Pilcher, Jennifer Day, and Tian Guo. 2020. "Great Lakes Region Acidification Research." In *NOAA Ocean, Coastal, and Great Lakes Acidification Research Plan: 2020–2029*, 101–8. https://oceanacidification.noaa.gov/sites/oap-redesign/11.%20Great%20Lakes.pdf.
Salehi, Ali Akbar, Mohammad Ghannadi-Maragheh, Meisam Torab-Mostaedi, Rezvan Torkaman, and Mehdi Asadollahzadeh. 2020. "A Review on the Water-Energy Nexus for Drinking Water Production from Humid Air." *Renewable and Sustainable Energy Reviews* 120:109627. https://doi.org/10.1016/j.rser.2019.109627.
Sanderson, Matthew R., Burke Griggs, and Jacob A. Miller-Klugesherz. 2020. "Farmers Are Depleting the Ogallala Aquifer Because the Government Pays Them to Do It." The Conversation, November 9, 2020. https://theconversation.com/farmers-are-depleting-the-ogallala-aquifer-because-the-government-pays-them-to-do-it-145501.
Scott, Michon. 2019. "National Climate Assessment: Great Plains' Ogallala Aquifer Drying Out." Climate.gov, NOAA (National Oceanic and Atmospheric Administration). February 19, 2019. https://www.climate.gov/news-features/featured-images/national-climate-assessment-great-plains%E2%80%99-ogallala-aquifer-drying-out.
Smith, Hayley. 2022. "Unprecedented Water Restrictions Hit Southern California Today: What They Mean to You." *Los Angeles Times*, June 1, 2022. https://www.latimes.com/california/story/2022-06-01/southern-california-new-drought-rules-june-2022.

Spiegel, Jan Ellen. 2021. "Climate Challenges Mount for California Agriculture." Yale Climate Connections. October 29, 2021. https://yaleclimateconnections.org/2021/10/climate-challenges-mount-for-california-agriculture/.

Stevens, Matt. 2015. "Water, 2015, California: The No-Good, Very Bad Year—Now, 'Pray for Rain.'" *Los Angeles Times*, September 29, 2015. https://www.latimes.com/local/lanow/la-me-ln-water-year-20150929-story.html.

Tandon, Ayesha. 2021. "Climate Change Has Driven 16% Drop in 'Snow Meltwater' from Asia's High Mountains." Carbon Brief. June 24, 2021. https://www.carbonbrief.org/climate-change-has-driven-16-drop-in-snow-meltwater-from-asias-high-mountains/.

UNICEF (The United Nations Children's Fund). 2021. "Water and the Global Climate Crisis: 10 Things You Should Know." March 18, 2021. https://www.unicef.org/stories/water-and-climate-change-10-things-you-should-know.

Uri, Emily. 2021. "Sea Level Rise Is Killing Trees along the Atlantic Coast, Creating 'Ghost Forests' That Are Visible from Space." The Conversation, April 6, 2021. https://theconversation.com/sea-level-rise-is-killing-trees-along-the-atlantic-coast-creating-ghost-forests-that-are-visible-from-space-147971.

USDA (US Department of Agriculture). 2015. *USDA Expands Investment in Water Conservation and Improvement in Nation's Largest Aquifer*. Natural Resources Conservation Service (NRCS). Accessed November 13, 2023. https://www.usda.gov/media/press-releases/2015/11/09/usda-expands-investment-water-conservation-and-improvement-nations,

US Global Change Research Program. 2014. "Water Supply." *Third National Climate Assessment*. https://nca2014.globalchange.gov/highlights/report-findings/water-supply.

Warka Water. n.d. Home page. Warka Water. Accessed May 23, 2022. https://www.warkawater.org.

WPS (Water, Peace, and Security Partnership). n.d. "About the Water, Peace, and Security Partnership." Accessed March 15, 2022. https://www.waterpeacesecurity.org/info/our-approach.

Space

CLIMATE REFUGEES

Ajibade, Idowu Jola, and A. R. Siders, eds. 2022. *Global Views on Climate Relocation and Social Justice: Navigating Retreat*. New York: Routledge.

Baker, Luke. 2020. "More than 1 Billion People Face Displacement by 2050—Report." Reuters, September 15, 2020. https://www.reuters.com/article/ecology-global-risks/more-than-1-billion-people-face-displacement-by-2050-report-idUSKBN2600K4.

Burnett, H. Sterling. 2021. "The Myth of Climate Refugees." Opinion, Townhall, April 10, 2021. https://townhall.com/columnists/hsterling burnett/2021/04/10/the-myth-of-climate-refugees-n2587694.

Climate Refugees. n.d. "Frontlines: Actions on Climate Displacement." Accessed November 22, 2021. https://www.climate-refugees.org/frontlinesactions.

Coastal Flood Resilience Project. 2022. *Proposed National Policies to Support Relocation of Communities as Sea Level Rises.* White paper, March 16, 2022. https://www.cfrp.info/.

Cornell University. 2017. "Rising Seas Could Result in 2 Billion Refugees by 2100." ScienceDaily, June 26, 2017. https://www.sciencedaily.com/releases/2017/06/170626105746.htm.

Ebbitt, Kathleen. 2015. "5 Facts on Climate Refugees—and Why You Should Care." Global Citizen. June 8, 2015. https://www.globalcitizen.org/en/content/5-facts-on-climate-refugees-and-why-you-should-car/.

Gaynor, Tim. 2020. "Climate Change Is the Defining Crisis of Our Time and It Particularly Impacts the Displaced." UNHCR the UN Refugee Agency. November 30, 2020. https://www.unhcr.org/news/stories/climate-change-defining-crisis-our-time-and-it-particularly-impacts-displaced.

Goodwin-Gill, Guy S., and Jane McAdam. 2017. *Climate Change, Disasters and Displacement.* Geneva, Switzerland: UN Refugee Agency, United Nations High Commissioner for Refugees (UNHCR). https://www.unhcr.org/596f25467.pdf.

Hughes, Rebecca Ann. 2021. "In Italy, 3D Printers Are Making Eco-Friendly Emergency Housing." *Forbes*, October 30, 2021. https://www.forbes.com/sites/rebeccahughes/2021/10/30/in-italy-3d-printers-are-making-eco-friendly-emergency-housing.

Jaffery, Rabiya. 2021. "Solar Power Changed Syrian Refugees' Lives in Jordan—and They Want More." Climate Home News, December 23, 2021. https://www.climatechangenews.com/2021/12/23/solar-power-changed-syrian-refugees-lives-jordan-want/.

Kroeger, Kyle. 2022. "13 Things to Know about Climate Refugees." *The Impact Investor* (blog), updated January 26, 2022. https://theimpactinvestor.com/climate-refugees/.

Lustgarten, Abrahm. 2020a. "How Climate Migration Will Reshape America." *New York Times Magazine*, September 15, 2020. https://www

.nytimes.com/interactive/2020/09/15/magazine/climate-crisis-migration-america.html.

Lustgarten, Abrahm. 2020b. "Where Will Everyone Go?" ProPublica, July 23, 2020. https://features.propublica.org/climate-migration/model-how-climate-refugees-move-across-continents.

McCarthy, Joe. 2022. "How War Impacts Climate Change and the Environment." Global Citizen. April 6, 2022. https://www.globalcitizen.org/en/content/how-war-impacts-the-environment-and-climate-change/.

National Geographic Society. 2019. "Climate Refugees." NG Resource Library/Encyclopedic Entry. Last updated March 28, 2019. https://www.nationalgeographic.org/encyclopedia/climate-refugees/.

Perle, Lola. 2021. "Climate Refugees: A Growing Crisis." Organization for World Peace. October 30, 2021. https://theowp.org/reports/climate-refugees-a-growing-crisis/.

Prescia, Moriah. 2021. "The Immediate Threat of Climate Change in Pakistan." Climate Refugees. July 30, 2021. https://www.climate-refugees.org/spotlight/2021/7/30/pakistan.

Tower, Amali. 2021. "VP's 'Root Causes' and Central America Strategies Lack Specificity and Climate Responsibility." Climate Refugees. August 2, 2021. https://www.climate-refugees.org/perspectives/2021/8/2/rootcausesstrategy.

Wallace-Wells, David. 2020. *The Uninhabitable Earth: Life after Warming.* New York: Tim Duggan Books/Penguin Random House.

The White House. 2021. *Report on the Impact of Climate Change on Migration.* A Report by the White House, Washington, DC, October 2021. https://www.whitehouse.gov/wp-content/uploads/2021/10/Report-on-the-Impact-of-Climate-Change-on-Migration.pdf.

World Bank Group. 2018. "Climate Change Could Force over 140 Million to Migrate within Countries by 2050: World Bank Report." Press Release No. 2018/118/CCG. March 19, 2018. https://www.worldbank.org/en/news/press-release/2018/03/19/climate-change-could-force-over-140-million-to-migrate-within-countries-by-2050-world-bank-report.

CLIMATE HAVENS

Brentjens, Emma. 2022. "10 Cities That Could Grow as Climate Change Worsens." LeafScore. Updated May 6, 2022. https://www.leafscore.com/blog/10-cities-that-could-grow-as-climate-change-worsens/.

Brodwin, Erin. 2018. "The Best Countries to Escape the Worst Effects of Climate Change." Business Insider, January 10, 2018. https://www.businessinsider.com/best-countries-escape-climate-change-map-2018-1.

Broom, Douglas. 2021. "A Third of Humanity Could Be on the Move If Climate Change Isn't Curbed, Scientists Say." World Economic Forum. November 3, 2021. https://www.weforum.org/agenda/2021/11/climate-change-rising-temperatures-may-force-humans-move/.

Holland, Oscar. 2017. "What Traditional Buildings Can Teach Architects about Sustainability." CNN, December 27, 2017. https://www.cnn.com/style/article/vernacular-architecture-sustainability/index.html.

Jacobson, Lindsey. 2022. "Americans Are Fleeing Climate Change — Here's Where They Can Go." CNBC, April 21, 2022. https://www.cnbc.com/2022/04/21/climate-change-encourages-homeowners-to-reconsider-legacy-cities.html.

Kaufman, Lisbeth. 2021. "What Is a Climate Haven? And Why I'm Moving to One." Climate Conscious, Medium, May 4, 2021. https://medium.com/climate-conscious/what-is-a-climate-haven-4f0efa2c7cbe.

Mercado, Angely. 2021. "Ancient Architecture Might Be Key to Creating Climate-Resilient Buildings." *Popular Science*, October 16, 2021. https://www.popsci.com/environment/vernacular-architecture-climate/.

Notre Dame Global Adaptation Initiative. 2023. "ND-GAIN Country Index." Notre Dame Research, University of Notre Dame. Accessed August 24, 2023. https://gain.nd.edu/our-work/country-index/rankings/.

Pierre-Louis, Kendra. 2019. "Want to Escape Global Warming? These Cities Promise Cool Relief." *New York Times*, April 15, 2019. https://www.nytimes.com/2019/04/15/climate/climate-migration-duluth.html.

Pogue, David. 2021. *How to Prepare for Climate Change: A Practical Guide to Surviving the Chaos*. New York: Simon and Schuster.

Royal Society of New Zealand Te Apārangi. 2016. "Climate Change Implications for New Zealand." https://www.royalsociety.org.nz/what-we-do/our-expert-advice/all-expert-advice-papers/climate-change-implications-for-new-zealand/.

Semuels, Alana. 2021. "Wildfires Are Getting Worse, So Why Is the U.S. Still Building Homes with Wood?" *Time*, June 2, 2021. https://time.com/6046368/wood-steel-houses-fires/.

Wallace-Wells, David. 2020. *The Uninhabitable Earth: Life after Warming*. New York: Tim Duggan Books/Penguin Random House.

Xu, Chi, Timothy A. Kohler, Timothy M. Lenton, Jens-Christian Svenning, and Marten Scheffer. 2020. "Future of the Human Climate Niche." *Proceedings of the National Academy of Sciences* (*PNAS*)117, no. 21: 11350–55. https://doi.org/10.1073/pnas.1910114117.

Yoder, Kate. 2021. "Fleeing Global Warming? 'Climate Havens' Aren't Ready for You Yet." Grist, December 7, 2021. https://grist.org/migration/fleeing-global-warming-climate-havens-arent-ready-for-you-yet/.

GREEN CITIES

Bales, Roger. 2022. "California Is about to Test Its First Solar Canals." *Smithsonian Magazine*, February 25, 2022. https://www.smithsonianmag.com/innovation/california-is-about-to-test-its-first-solar-canals-180979637/.

EPA (US Environmental Protection Agency). 2022. "Heat Island Effect." Last updated June 6, 2022. https://www.epa.gov/heatislands.

Global Cool Cities Alliance. 2012. "A Practical Guide to Cool Roofs and Cool Pavements." Cool Roofs and Cool Pavements Toolkit. January 2012. https://coolrooftoolkit.org/wp-content/pdfs/CoolRoofToolkit_Full.pdf.

Grossman, David. 2019. "Climate Change Could Wreck a Quarter of Steel Bridges in 21 Years." *Popular Mechanics*, October 25, 2019. https://www.popularmechanics.com/technology/infrastructure/a29579577/climate-changes-bridges/.

New York State Governor's Office of Storm Recovery. 2020. "Living Breakwaters Project Background and Design." Office of Resilient Homes and Communities. https://stormrecovery.ny.gov/living-breakwaters-project-background-and-design.

Short, John Rennie. 2021. "Cities Worldwide Aren't Adapting to Climate Change Quickly Enough." The Conversation, October 20, 2021. https://theconversation.com/cities-worldwide-arent-adapting-to-climate-change-quickly-enough-169984.

Sisson, Patrick. 2017. "Cycling Success: 10 U.S. Cities Pushing Biking Forward." Curbed, April 18, 2017. https://archive.curbed.com/2017/4/18/15333796/best-cities-bike-commute-us-cycling.

UNEP (United Nations Environment Programme). 2021. *A Practical Guide to Climate-Resilient Buildings and Communities.* Nairobi: United Nations Environment Programme. https://www.unep.org/resources/practical-guide-climate-resilient-buildings.

UNEP (United Nations Environment Programme). n.d. "Share the Road." Accessed May 23, 2022. https://www.unep.org/explore-topics/transport/what-we-do/share-road.

UN-Habitat (United Nations Human Settlements Programme). 2022. "UN-Habitat and Partners Unveil OCEANIX Busan, the World's First Prototype Floating City." April 27, 2022. https://unhabitat.org/un-habitat-and-partners-unveil-oceanix-busan-the-worlds-first-prototype-floating-city.

Ziezulewicz, Geoff. 2022. "USS Kitty Hawk Headed for the Scrapyard." *Navy Times*, January 18, 2022. https://www.navytimes.com/news/your-navy/2022/01/18/uss-kitty-hawk-headed-for-the-scrapyard/.

HEALTH

Allen, Joseph G., Piers MacNaughton, Usha Satish, Suresh Santanam, Jose Vallarino, and John D. Spengler. 2016. "Associations of Cognitive Function Scores with Carbon Dioxide, Ventilation, and Volatile Organic Compound Exposures in Office Workers: A Controlled Exposure Study of Green and Conventional Office Environments." *Environmental Health Perspectives* 124, no. 6. https://doi.org/10.1289/ehp.1510037.

CDC (Centers for Disease Control and Prevention). 2020. *Preparing for the Regional Health Impacts of Climate Change in the United States*." Climate and Health Program, July 2020. https://www.cdc.gov/climateand health/docs/Health_Impacts_Climate_Change-508_final.pdf.

CDC (Centers for Disease Control and Prevention). 2022a. "Climate Effects on Health." National Center for Environmental Health. Last reviewed April 25, 2022. https://www.cdc.gov/climateandhealth/effects/default .htm.

CDC (Centers for Disease Control and Prevention). 2022b. "Emergency Preparedness for People with Chronic Diseases." National Center for Chronic Disease Prevention and Health Promotion. Updated April 6, 2022. https://www.cdc.gov/chronicdisease/resources/infographic /emergency.htm.

Collier, Aram. n.d. "Melting Ice Reveals Secrets about Human History." *The Nature of Things*, CBC. Accessed July 23, 2022. https://www.cbc.ca /natureofthings/features/melting-ice-reveals-secrets-about -human-history.

Dey, Tanujit, Antonella Zanobetti, and Clas Linnman. 2023. "The Risk of Being Bitten by a Dog Is Higher on Hot, Sunny, and Smoggy Days." *Scientific Reports* 13, 8749 (2023). https://doi.org/10.1038/s41598-023 -35115-6.

Ebi, K. L., J. M. Balbus, G. Luber, A. Bole, A. Crimmins, G. Glass, S. Saha, M. M. Shimamoto, J. Trtanj, and J. L. White-Newsome. 2018. "Human Health." In *Fourth National Climate Assessment. Vol 2, Impacts, Risks, and Adaptation in the United States: Fourth National Climate Assessment*, edited by D. R. Reidmiller, C. W. Avery, D. R. Easterling, K. E. Kunkel, K. L. M. Lewis, T. K. Maycock, and B. C. Stewart, 2:539–71. Washington, DC: US Global Change Research Program. https://nca2018 .globalchange.gov/chapter/14/.

Gorvett, Zarla. 2020. "The Troubling Ways a Heatwave Can Warp Your Mind." BBC Future, August 17, 2020. https://www.bbc.com/future /article/20200817-the-sinister-ways-heatwaves-warp-the-mind.

Harp, Ryan D., and Kristopher V. Karnauskas. 2020. "Global Warming to Increase Violent Crime in the United States." *Environmental*

Research Letters 15, no. 3: 034039. https://iopscience.iop.org/article/10.1088/1748-9326/ab6b37.

Lowe, Robert J., Gesche M. Huebner, and Tadj Oreszczyn. 2018. "Possible Future Impacts of Elevated Levels of Atmospheric CO_2 on Human Cognitive Performance and on the Design and Operation of Ventilation Systems in Buildings." *Building Services Engineering Research and Technology* 39, no. 6: 698–711. https://doi.org/10.1177/0143624418790129.

Morganstein, Joshua C. 2019. "How Extreme Weather Events Affect Mental Health." American Psychiatric Association. November 2019. https://psychiatry.org/patients-families/climate-change-and-mental-health-connections/affects-on-mental-health.

ReadyCare. n.d. "How a FRIO Works." FRIO® Insulin Cooling Case. Accessed July 3, 2023. https://www.frioinsulincoolingcase.com/how-a-frio-works/#.

Sommer, Lauren. 2021. "Climate Change Is the Greatest Threat to Public Health, Top Medical Journals Warn." *Morning Edition*, NPR, September 7, 2021. https://www.npr.org/2021/09/07/1034670549/climate-change-is-the-greatest-threat-to-public-health-top-medical-journals-warn.

United States Global Change Research Program. 2016a. "Farm to Table: The Potential Interactions of Rising CO2 and Climate Change on Food Safety and Nutrition." U.S. Climate Resilience Toolkit. Last modified November 16, 2016. https://toolkit.climate.gov/image/2005.

United States Global Change Research Program. 2016b. "Food- and Water-Related Threats." U.S. Climate Resilience Toolkit. Last modified November 16, 2016. https://toolkit.climate.gov/topics/human-health/changing-ecosystems.

Wallace-Wells, David. 2020. *The Uninhabitable Earth: Life after Warming.* New York: Tim Duggan Books/Penguin Random House.

Weiss, Sabrina. 2022. "How Heat Waves Are Messing Up Your Sleep." *Glamour*, July 15, 2022. https://www.glamour.com/story/how-heat-waves-are-messing-up-your-sleep.

Zhang, Xin, Xi Chen, and Xiaobo Zhang. 2018. "The Impact of Exposure to Air Pollution on Cognitive Performance." *Proceedings of the National Academy of Sciences* (*PNAS*) 115, no. 37: 9192–97. https://doi.org/10.1073/pnas.1809474115.

Zhang, Yingxiao, and Allison L. Steiner. 2022. "Projected Climate-Driven Changes in Pollen Emission Season Length and Magnitude over the Continental United States." *Nature Communications* 13:1234. https://www.nature.com/articles/s41467-022-28764-0.

NATURE ON THE MOVE

Biello, David. 2008. "Climbing Trees: Plants Move Uphill as World Warms." *Scientific American*, June 26, 2008. https://www.scientificamerican.com/article/climbing-trees-plants-move-uphill-as-climate-changes/.

Kerlin, Kat. 2017. "Earth's Oldest Trees in Climate-Induced Race up the Tree Line." UC Davis News. September 13, 2017. https://www.ucdavis.edu/news/earths-oldest-trees-climate-induced-race-tree-line.

Kolbert, Elizabeth. 2014. *The Sixth Extinction: An Unnatural History*. New York: Henry Holt.

Kummel, Miroslav, Alison McGarigal, Michelle Kummel, Carol Earnest, and Molly Feiden. 2021. "Tree Establishment and Growth Drive Treeline Advance and Change Treeline Form on Pikes Peak (Colorado) in Response to Recent Anthropogenic Warming." *Canadian Journal of Forest Research* 51, no. 10 (February 14, 2021). https://doi.org/10.1139/cjfr-2019-0358.

National Park Service. 2008. "USGS Maps Show Potential Non-native Python Habitat along 3 US Coasts." Everglades National Park, Florida, news release, February 20, 2008; last updated April 14, 2015. https://www.nps.gov/ever/learn/news/usgs-maps-show-potential-non-native-python-habitat-along-3-us-coasts.htm.

PBS. 2018. "How the Pika Adapts to Climate Change." YouTube video, May 31, 2018. https://www.youtube.com/watch?v=IF8ym4g2SCU.

Steinauer-Scudder, Chelsea, and Jeremy Seifert. 2021. "They Carry Us With Them: The Great Tree Migration." Multimedia article, directed by Jeremy Seifert, produced by Emmanuel Vaughan-Lee. *Emergence Magazine*, November 4, 2021. https://emergencemagazine.org/feature/they-carry-us-with-them/.

Wired. 2021. "How Animals Are Rapidly Evolving Because of Climate Change." YouTube video, November 29, 2021. https://www.youtube.com/watch?v=uMxsfImpaYM.

THE BIOSPHERE

Bar-On, Yinon M., Rob Phillips, and Ron Milo. 2018. "The Biomass Distribution on Earth." *Proceedings of the National Academy of Sciences* (*PNAS*) 115, no 25: 6506–11. https://doi.org/10.1073/pnas.1711842115.

Boyce, James K. 2021. "The Environmental Cost of Inequality." *Scientific American*, special ed., Fall 2021, 76–81.

Bruckner, Monica. 2022. "Endoliths—Microbes Living within Rocks." Microbial Life Educational Resources, SERC, Carleton College. Last modified June 8, 2022. https://serc.carleton.edu/microbelife/extreme/endoliths/index.html.

Carrington, Damian. 2018. "Humans Just 0.01% of All Life but Have Destroyed 83% of Wild Mammals—Study." *Guardian*, May 21, 2018. https://www.theguardian.com/environment/2018/may/21/human-race-just-001-of-all-life-but-has-destroyed-over-80-of-wild-mammals-study.

Chivian, Eric, and Aaron Bernstein. 2010. *How Our Health Depends on Biodiversity*. Center for Health and the Global Environment, Harvard Medical School. https://www.bu.edu/sph/files/2012/12/Chivian_and_Bernstein_2010_How_our_Health_Depends_on_Biodiversity.pdf.

Connellan, Shannon. 2022. "23 Climate Change Documentaries You Need to Watch Because This Planet is NOT Fine." Mashable, August 10, 2022. https://mashable.com/article/best-climate-change-documentaries.

Diamond, Jared. 2011. *Collapse: How Societies Choose to Fail or Succeed*. Rev. ed. New York: Penguin.

Dunn, Rob. 2021. *A Natural History of the Future: What the Laws of Biology Tell Us about the Destiny of the Human Species*. New York: Basic Books.

Fullerton, John. 2015. *Regenerative Capitalism: How Universal Principles and Patterns Will Shape Our New Economy*. Stonington, CT: Capital Institute, April 2015. https://capitalinstitute.org/wp-content/uploads/2015/04/2015-Regenerative-Capitalism-4-20-15-final.pdf.

"IPAT." 2022. Oxford Reference, Oxford University Press. Accessed June 17, 2022. https://www.oxfordreference.com/view/10.1093/oi/authority.20110803100010550.

Kirschke-Schwartz, Evie. 2016. "Wild Laws: China and Its Role in Illicit Wildlife Trade." *New Security Beat* (blog), Wilson Center, September 12, 2016. https://www.newsecuritybeat.org/2016/09/wild-laws-china-role-wildlife-trafficking/.

Mills, J. A. 2015. *The Blood of the Tiger: A Story of Conspiracy, Greed, and the Battle to Save a Magnificent Species*. Boston: Beacon.

National Geographic Society. n.d. "Biosphere." Resource Library, Encyclopedic Entry. Accessed February 4, 2022. https://www.nationalgeographic.org/encyclopedia/biosphere/.

Nuwer, Rachel Love. 2018. *Poached: Inside the Dark World of Wildlife Trafficking*. Boston: Da Capo.

Parry, Wynne. 2013. "Sky-High Microbes: How Far Up Can Life Exist?" Live Science, November 13, 2013. https://www.livescience.com/41173-sky-high-microbes-how-far-up-can-life-exist.html.

Stokel-Walker, Chris. 2021. "Regenerative Capitalism Is Coming—and This Is What It Means to Your Industry." Neste, November 10, 2021. https://journeytozerostories.neste.com/sustainability/regenerative-capitalism-coming-and-what-it-means-your-industry.

Surma, Katie. 2021. "Ecuador's High Court Affirms Constitutional Protections for the Rights of Nature in a Landmark Decision." Inside Climate News, December 3, 2021. https://insideclimatenews.org/news/03122021/ecuador-rights-of-nature/.

Wilson, Edward O. 2016. *Half-Earth: Our Planet's Fight for Life*. New York: Liveright/W. W. Norton.

Wilson, Joseph. 2022. " Spain Gives Personhood Status to Mar Menor Salt-Water Lagoon." *US News and World Report*, September 21, 2022. https://www.usnews.com/news/world/articles/2022-09-21/spain-grants-personhood-status-to-mar-menor-lagoon.

The Heart of the Matter

IPCC (Intergovernmental Panel on Climate Change). 2021a. *Climate Change 2021: The Physical Science Basis; Contribution of Working Group I to the Sixth Assessment Report of the Intergovernmental Panel on Climate Change*, edited by Valérie Masson-Delmotte, Panmao Zhai, Anna Pirani, Sarah L. Connors, Clotilde Péan, Sophie Berger, Nada Caud, et al. Cambridge: Cambridge University Press. https://www.ipcc.ch/report/ar6/wg1/#SPM.

IPCC (Intergovernmental Panel on Climate Change). 2021b. "Summary for Policymakers." In *Climate Change 2021: The Physical Science Basis; Contribution of Working Group I to the Sixth Assessment Report of the Intergovernmental Panel on Climate Change*, edited by Valérie Masson-Delmotte, Panmao Zhai, Anna Pirani, Sarah L. Connors, Clotilde Péan, Sophie Berger, Nada Caud, et al., 3–32. Cambridge: Cambridge University Press. https://www.ipcc.ch/report/sixth-assessment-report-working-group-i/.

IPCC (Intergovernmental Panel on Climate Change). 2022. *Climate Change 2022: Impacts, Adaptation and Vulnerability, Working Group II Contribution to the Sixth Assessment Report of the Intergovernmental Panel on Climate Change*, edited by H.-O. Pörtner, D. C. Roberts, M. M. B. Tignor, E. Poloczanska, K. Mintenbeck, A. Alegría, M. Craig, et al. Cambridge: Cambridge University Press. https://doi.org/10.1017/9781009325844.

New Ideas

Bever, Fred. 2021. "'Run the Oil Industry in Reverse': Fighting Climate Change by Farming Kelp." Maine Public, NPR, March 1, 2021. https://www.npr.org/2021/03/01/970670565/run-the-oil-industry-in-reverse-fighting-climate-change-by-farming-kelp.

Hawken, Paul, ed. 2017. *Drawdown: The Most Comprehensive Plan Ever Proposed to Reverse Global Warming.* New York: Penguin Books.
The International "4 per 1000" Initiative: Soils for Food Security and Climate. n.d. Accessed July 26, 2022. https://4p1000.org/?lang=en.
Kolbert, Elizabeth. 2021. *Under a White Sky: The Nature of the Future.* New York: Crown.
Ledig, F. Thomas, and J. H. Kitzmiller. 1992. "Genetic Strategies for Reforestation in the Face of Global Climate Change." *Forest Ecology and Management* 50:153–69. https://www.fs.usda.gov/psw/publications/ledig/psw_1992_ledig017.pdf.
Little, Jane Braxton. 2021. "Firefighting Robots Go Autonomous." *Scientific American*, October 29, 2021. https://www.scientificamerican.com/article/firefighting-robots-go-autonomous/.
Markham, Lauren. 2021. "The (Very Slow) Race to Move Forests in Time to Save Them." *Wired*, October 9, 2021. https://www.wired.com/story/the-very-slow-race-to-move-forests-in-time-to-save-them/.
MIT. 2022. "In Case of Climate Emergency: Deploying Space Bubbles to Block Out the Sun." SciTechDaily, July 16, 2022. https://scitechdaily.com/in-case-of-climate-emergency-deploying-space-bubbles-to-block-out-the-sun/.
Peters, Adele. 2017. "These Tree-Planting Drones Are about to Start an Entire Forest from the Sky." Fast Company, August 10, 2017. https://www.fastcompany.com/40450262/these-tree-planting-drones-are-about-to-fire-a-million-seeds-to-re-grow-a-forest.
Robinson, Kim Stanley. 2020. *The Ministry for the Future: A Novel.* New York: Orbit.
Rumpel, Cornelia, Farshad Amiraslani, Claire Chenu, Magaly Garcia Cardenas, Martin Kaonga, Lydie-Stella Koutika, Jagdish Ladha, et al. 2020. "The 4p1000 Initiative: Opportunities, Limitations and Challenges for Implementing Soil Organic Carbon Sequestration as a Sustainable Development Strategy." *Ambio* 49, no. 1: 350–60. https://doi.org/10.1007/s13280-019-01165-2.
Smedley, Tim. 2019. "How Artificially Brightened Clouds Could Stop Climate Change." Science and Environment, BBC, February 2, 2019. https://www.bbc.com/future/article/20190220-how-artificially-brightened-clouds-could-stop-climate-change.
Wood, Daniel. 2014. "Space-Based Solar Power." Energy.gov, Department of Energy. March 6, 2014. https://www.energy.gov/articles/space-based-solar-power.

New Developments

CNN. 2023. "Scientists Warn Antarctic Sea Ice Record Low." Video, March 1, 2023. https://www.cnn.com/videos/world/2023/03/01/antarctic-sea-ice-record-climate-change-weir-vpx.cnn.

Douglas, Leah. 2023. "Insight: Lab-grown Meat Moves Closer to American Dinner Plates." Reuters, January 23, 2023. https://www.reuters.com/business/retail-consumer/lab-grown-meat-moves-closer-american-dinner-plates-2023-01-23/.

IPCC (Intergovernmental Panel on Climate Change). 2023. *AR6 Synthesis Report: Climate Change 2023*. https://www.ipcc.ch/report/ar6/syr/.

Lea, Robert. 2023. "Doomsday Glacier Melting in Antarctica Means Terrible News for Global Sea Level Rise." Space.com, February 17, 2023. https://www.space.com/antarctica-doomsday-glacier-melts-in-treacherous-ways.

McLay, Charlotte-Graham. 2023. "Auckland Floods: City Begins Clean-up after 'Biggest Climate Event' in New Zealand's History." *Guardian*, January 31, 2023. https://www.theguardian.com/world/2023/feb/01/auckland-floods-biggest-climate-event-new-zealand-history-flooding.

Ramirez, Rachel. 2023a. "Large Glacier Near Seattle Has 'Completely Disappeared,' Says Researcher Who Has Tracked It for Years." CNN, February 1, 2023. Updated February 2, 2023. https://www.cnn.com/2023/02/01/us/washington-hinman-glacier-disappear-climate/index.html.

Ramirez, Rachel. 2023b. "An 'Inland Tsunami': 15 Million People Are at Risk from Catastrophic Glacial Lake Outbursts, Researchers Find." CNN, February 7, 2023. https://www.cnn.com/2023/02/07/world/glacial-lake-outburst-risk-climate-scn/index.html.

Ripple, William J., Christopher Wolf, Timothy M. Lenton, Jillian W. Gregg, Susan M. Natali, Philip B. Duffy, Johan Rockstrom, and Hans Joachim Schellnhuber. 2023. "Many Risky Feedback Loops Amplify the Need for Climate Action." One Earth. https://doi.org/10.1016/j.oneear.2023.01.004.

Bug-Out Bags

Mahoney, Doug, and Joshua Lyon. 2022. "The Best Gear for Your Bug-Out Bag." *New York Times*, updated January 12, 2022. https://www.nytimes.com/wirecutter/reviews/best-gear-for-your-bug-out-bag/.

Outside Interactive. 2022. "How to Build the Ultimate Bug-Out Bag." Outside Online, June 21, 2022. https://www.outsideonline.com/outdoor-gear/how-to-build-the-ultimate-bug-out-bag/.

Stout, James. 2021. "29 Items Your Ultimate Bug-Out Bag Should Have." Gear Patrol, updated August 30, 2021. https://www.gearpatrol.com/outdoors/a263353/bugout-bag/.

US Department of Homeland Security. 2022. "Build a Kit." Ready.gov. Last updated May 10, 2022. https://www.ready.gov/kit.

ABOUT THE AUTHORS

ORRIN PILKEY is Emeritus James B. Duke Professor of Earth and Ocean Sciences, Duke University. Once a deep-sea sedimentologist specializing in abyssal plains, he became a coastal geologist after Hurricane Camille destroyed his parents' retirement home in Waveland, Mississippi, in 1969.

CHARLES PILKEY is a former geologist turned freelance sculptor, writer, and illustrator. During the period of his "Babylonian captivity," he worked for an oil company in the western United States and Canada. He sometimes writes books and editorials with his father.

LINDA PILKEY-JARVIS, geologist, daughter of Orrin Pilkey, has coauthored two books with him: *Retreat from a Rising Sea* and *Useless Arithmetic*. Linda lives in the Pacific Northwest, and her many years of applying environmental policy at the state level give her an understanding of how government can shape solutions to problems.

NORMA LONGO, longtime technical associate of Orrin Pilkey, is a photographer, geologist, media researcher, and copyeditor. Coauthor with Pilkey of numerous articles and some books, including *Vanishing Sands: Losing Beaches to Mining*.

KEITH PILKEY, an administrative law judge in the Social Security system, son of Orrin Pilkey, has been involved in Orrin's research activities for years and has coauthored several books with him, including *Vanishing Sands*, *Sea Level Rise*, *Global Climate Change: A Primer*, and *Retreat from a Rising Sea*.

FRED DODSON, a real estate developer, CEO, and executive vice president of DreamKey Partners, Inc., manages the organization's affordable housing development activities. His decades of real estate development experience give him a helpful view of the climate haven problems.

HANNAH HAYES is a scholar of climate justice, decarbonization, sustainable development, and disaster capitalism. Holding a BA from the University of Tennessee and an MS from Trinity College Dublin, she coauthored *Vanishing Sands* with Orrin.

INDEX

Page references in *italics* indicate illustrations and *t* indicates a table.